# UNDERSTANDING THE NEW US DEFENSE POLICY

*Through the Speeches of*

## ROBERT M. GATES, SECRETARY OF DEFENSE

SPEECHES & REMARKS DECEMBER 18, 2006 TO FEBRUARY 10, 2008
as Released by the U.S. Department of Defense

Rockville, Maryland
2008

*Understanding US Defense Policy Through the Speeches of Robert M. Gates, Secretary of Defense - Speeches & Remarks December 18, 2006 to February 10, 2008* by **Robert M. Gates as Released to the Public by the US Department of Defense** in its current format, copyright © Arc Manor 2008. The contents themselves have been released to the pubic domain by the Department of Defense. The index is copyrighted © Arc Manor 2008.

Arc Manor, Arc Manor Classic Reprints, Manor Classics, TARK Classic Fiction and the Arc Manor logo are trademarks or registered trademarks of Arc Manor Publishers, Rockville, Maryland. All other trademarks are properties of their respective owners.

This book is presented as is, without any warranties (implied or otherwise) as to the accuracy of the production, text or translation. The publisher does not take responsibility for any typesetting, formatting, translation or other errors which may have occurred during the production of this book.

ISBN: 978-1-60450-103-2

Library of Congress Cataloging-in-Publication Data

Gates, Robert Michael, 1943-
  Understanding the new US defense policy through the speeches of Robert M. Gates, Secretary of Defense : speeches & remarks December 18, 2006 to February 10, 2008 as released by the US Department of Defense.
    p. cm.
  Includes index.
  ISBN 978-1-60450-103-2 (alk. paper)
  1. United States--Military policy--21st century. 2. United States. Dept. of Defense. I. Title.
  UA23.G525 2008
  355'.033573--dc22
                    2008007602

Information and illustrations (including photographs of Dr. Gates) contained in this book has been released to the public by the US Department of Defense. Arc Manor, LLC is not associated with the US Department of Defense or any other agency of the US Federal Government. Arc Manor has compiled the information contained in this book through many means which include (but are not limited to) US Government documents and web-sites and other information released through the Freedom of Information Act (FOIA) and otherwise made public by relevant federal departments/agencies.

Published by Arc Manor
P. O. Box 10339
Rockville, MD 20849-0339
www.ArcManor.com
Printed in the United States of America/United Kingdom

## Contents

SECRETARY GATES' SWEARING-IN REMARKS
*Pentagon Auditorium, The Pentagon, Monday, December 18, 2006*     7

PRESS AVAILABILITY WITH SECRETARY OF STATE CONDOLEEZZA RICE
*White House, Washington, DC, Thursday, January 11, 2007*     9

TESTIMONY ON IRAQ TO THE HOUSE ARMED SERVICES COMMITTEE
*Washington, DC, Thursday, January 11, 2007*     12

STATEMENT ON IRAQ TO THE SENATE ARMED SERVICE COMMITTEE
*Washington, DC, Friday, January 12, 2007*     15

POSTURE STATEMENT TO THE SENATE ARMED SERVICES COMMITTEE
*Washington, DC, Tuesday, February 06, 2007*     18

MUNICH CONFERENCE ON SECURITY POLICY
*Munich, Germany, Sunday, February 11, 2007*     25

U.S. SENATE COMMITTEE ON APPROPRIATIONS
*Washington, DC, Tuesday, February 27, 2007*     31

SENIOR LEADERSHIP MEETING OF THE NATIONAL GUARD
*National Guard Bureau, Washington, DC, Tuesday, February 27, 2007*     35

NCO BREAKFAST
*Fort Belvoir, VA, Thursday, March 01, 2007*     39

U.S. SENATE YOUTH PROGRAM
*Pentagon Auditorium, Washington, DC, Friday, March 09, 2007*     42

U.S. NORTHERN COMMAND CHANGE OF COMMAND CEREMONY
*Peterson Air Force Base Colorado Springs, Colorado, Friday, March 23, 2007*     45

U.S. PACIFIC COMMAND ASSUMPTION OF COMMAND CEREMONY
*Camp H.M. Smith, Hawaii, Monday, March 26, 2007*     49

AMERICAN-TURKISH COUNCIL
*Washington, DC, Tuesday, March 27, 2007*     52

HOUSE APPROPRIATIONS COMMITTEE—DEFENSE
*Washington, DC, Thursday, March 29, 2007*     58

ARMY CHIEF OF STAFF CHANGE OF RESPONSIBILITY CEREMONY
*Fort Myer, Virginia, Tuesday, April 10, 2007*     65

AMERICAN CHAMBER OF COMMERCE OF CAIRO
*Cairo, Egypt, Wednesday, April 18, 2007*   69

NAVY FLAG OFFICERS CONFERENCE
*U.S. Naval Academy, Wednesday, April 25, 2007*   74

GREATER DALLAS CHAMBER OF COMMERCE
*Dallas, Texas, Thursday, May 03, 2007*   77

SENATE APPROPRIATIONS COMMITTEE
*106 Dirksen Senate Office Building, Washington, DC, Wednesday, May 09, 2007* 84

TEAM AMERICA ROCKETRY CHALLENGE
*The Plains, Virginia, Saturday, May 19, 2007*   91

COLLEGE OF WILLIAM AND MARY COMMENCEMENT
*Williamsburg, Virginia, Sunday, May 20, 2007*   93

UNITED STATES NAVAL ACADEMY COMMENCEMENT
*Annapolis, Maryland, Friday, May 25, 2007*   99

AIR FORCE ACADEMY COMMENCEMENT
*Colorado Springs, Colorado, Wednesday, May 30, 2007*   105

INTERNATIONAL INSTITUTE FOR STRATEGIC STUDIES
*Singapore, Saturday, June 02, 2007*   111

COMMEMORATION OF THE 63RD ANNIVERSARY OF D-DAY
*Colleville-Sur-Mer, France, Wednesday, June 06, 2007*   120

U.S. SPECIAL OPERATIONS COMMAND CHANGE OF COMMAND CEREMONY
*Tampa Convention Center, Tampa, Florida, Monday, July 09, 2007*   123

MARINE CORPS ASSOCIATION ANNUAL DINNER
*Crystal Gateway Marriott Hotel, Arlington, VA, Wednesday, July 18, 2007*   127

FAREWELL TO ADMIRAL GIAMBASTIANI
*Annapolis, MD, Friday, July 27, 2007*   135

SEPTEMBER 11TH WREATH-LAYING CEREMONY
*Pentagon, 9/11 Cornerstone, Tuesday, September 11, 2007*   137

WORLD FORUM ON THE FUTURE OF DEMOCRACY
*Williamsburg, VA, Monday, September 17, 2007*   139

UNITED STATES AIR FORCE 60TH ANNIVERSARY
*Pentagon Courtyard, Tuesday, September 18, 2007*   146

POW/MIA RECOGNITION DAY
*Pentagon Parade Field, Friday, September 21, 2007*   14

SENATE APPROPRIATIONS COMMITTEE
*106 Dirksen Senate Office Building, Washington, DC, Wednesday, September 26, 2007*   151

FAREWELL CEREMONY FOR GENERAL PETER PACE
*Summerall Field, Fort Myer, VA, Monday, October 01, 2007*     154

ASSOCIATION OF THE UNITED STATES ARMY
*Washington, DC, Wednesday, October 10, 2007*     157

MILITARY ACADEMY OF THE GENERAL STAFF
*Moscow, Russia, Saturday, October 13, 2007*     164

JEWISH INSTITUTE FOR NATIONAL SECURITY AFFAIRS
*Arlington, VA, Monday, October 15, 2007*     169

STRATCOM ASSUMPTION OF COMMAND
*Offutt Air Force Base, NE, Wednesday, October 17, 2007*     176

CONFERENCE OF EUROPEAN ARMIES
*Heidelberg, Germany, Thursday, October 25, 2007*     179

GEORGE BUSH AWARD FOR EXCELLENCE IN PUBLIC SERVICE
*College Station, TX, Friday, October 26, 2007*     184

SOPHIA UNIVERSITY
*Tokyo, Japan, Friday, November 09, 2007*     191

MILITARY SPOUSE CAREER ADVANCEMENT INITIATIVE
*Pentagon Conference Center, Wednesday, November 14, 2007*     197

BOY SCOUTS OF AMERICA "CITIZEN OF THE YEAR" AWARD
*Washington, DC, Thursday, November 15, 2007*     199

BASE COMMUNITY COUNCIL
*Whiteman Air Force Base, Missouri, Tuesday, November 20, 2007*     202

GREATER KILLEEN CHAMBER OF COMMERCE
*Killeen, TX, Monday, November 26, 2007*     206

LANDON LECTURE
*Manhattan, Kansas, Monday, November 26, 2007*     210

MANAMA DIALOGUE
*Manama, Bahrain, Saturday, December 08, 2007*     222

HOUSE ARMED SERVICES COMMITTEE HEARING ON AFGHANISTAN
*Capitol Hill, Washington, DC, Tuesday, December 11, 2007*     230

23RD ANNUAL MARTIN LUTHER KING, JR. OBSERVANCE
*Washington D.C., Thursday, January 17, 2008*     233

SPACE AND NAVAL WARFARE CENTER MRAP FACILITY
*Charleston, SC, Friday, January 18, 2008*     235

AMERICA SUPPORTS YOU SUMMIT
*Pentagon, Friday, January 25, 2008*     237

Center for Strategic and International Studies
*Washington, DC, Saturday, January 26, 2008*     239

Senate Armed Services Committee Opening Remarks
*Washington, D.C., Wednesday, February 06, 2008*     246

House Armed Services Committee Opening Remarks
*Washington, D.C., Wednesday, February 06, 2008*     249

Appendix I Dr. Robert M. Gates     253

Appendix II Responsibilities of the Office of the Secretary of Defense     255

Appendix III – Organizational Chart     257

Appendix IV     259
  Munich Conference on Security Policy
  *Munich, Germany, Sunday, February 10, 2008*

Index     267

## Secretary Gates' Swearing-In Remarks

*Pentagon Auditorium, The Pentagon, Monday, December 18, 2006*

THANK you. Mr. President, I am deeply honored by the trust you have placed in me. You have asked for my candor and my honest counsel at this critical moment in our nation's history, and you will get both.

Mr. Vice President, thank you for administering the oath of office. I first worked closely with the Vice President when he was a very successful Secretary of Defense, and I hope some of that may rub off.

My sincere thanks to the members of the United States Congress who are here today. I appreciate the prompt and fair hearing that I received in the Senate and the confidence that senators have placed in me.

Chairman Pace, thank you. I look forward to working with you and the Joint staff.

To the service chiefs and the service staffs, to all the uniform military here today, I value your professionalism and your experience, and I will rely on your clear-eyed advice in the weeks and months ahead.

Finally, I want to thank Becky, my wife of 40 years; and my children, Eleanor and Brad, for their infinite patience. I want to thank other family and friends who are here, but single out one especially—my 93-year-old mother. She told me that if she could make it from Kansas to Texas A&M football games every fall, she certainly could be in Washington for this ceremony.

I, too, want to say a few words about my predecessor. Donald Rumsfeld has devoted decades of his life to public service. He cares deeply about our men and women in uniform, and the future of our country. I thank him for his long and distinguished service, and wish him and Joyce and their family all the best.

It is an honor to have the opportunity to work with the people in this Department, dedicated professionals whose overriding priority is the defense of our nation. Long ago, I learned something about leading large institutions: Leaders come and go, but the professionals endure long after the appointees are gone. The key to successful leadership in my view is to involve in the decision-making process early and often those who ultimately must carry out the decisions. I will do my best to do just that.

This Department, as always, is carrying on many different activities all at the same time. All are valuable, all are important. However, as I said in my confirmation hearings, Iraq is at the top of the list.

In the days since the Senate confirmed me, I have participated in most of the National Security Council meetings on Iraq, I have received a number of briefings here at the Department of Defense, and I have discussed the situation and way forward in Iraq in depth with the President.

I intend to travel quite soon to Iraq and meet with our military leaders and other personnel there. I look forward to hearing their honest assessments of the situation on the ground and to having the benefit of their advice—unvarnished and straight from the shoulder—on how to proceed in the weeks and months ahead.

Another pressing concern is Afghanistan. The progress made by the Afghan people over the past five years is at risk. The United States and its NATO allies have made a commitment to the Afghan people, and we intend to keep it. Afghanistan cannot be allowed to become a sanctuary for extremists again.

How we face these and other challenges in the region over the next two years will determine whether Iraq, Afghanistan, and other nations at a crossroads will pursue paths of gradual progress towards sustainable governments, which are allies in the global war on terrorism, or whether the forces of extremism and chaos will become ascendant.

All of us want to find a way to bring America's sons and daughters home again. But, as the President has made clear, we simply cannot afford to fail in the Middle East. Failure in Iraq at this juncture would be a calamity that would haunt our nation, impair our credibility, and endanger Americans for decades to come.

Finally, there is the matter of what is referred to as defense transformation. As I mentioned in my Senate testimony, I was impressed by how deployable our military has become since I last served in government.

Before he came to office, the President said that one of his top priorities was to help our military become more agile, more lethal, and more expeditionary. Much has been accomplished in this; much remains to be done. This remains a necessity and a priority.

I return to public service in the hope that I can make a difference at a time when our nation is facing daunting challenges and difficult choices. Mr. President, I thank you again for the opportunity to do that, and thank all of you for being here.

## Press Availability With Secretary of State Condoleezza Rice

*White House, Washington, DC, Thursday, January 11, 2007*

THANK you, Secretary Rice.

This afternoon, General Pace and I will appear before the House Armed Services Committee to discuss the military aspects of the Iraq strategy announced by the President last night. Tomorrow, we will appear before the Senate Armed Services Committee.

The security plan is designed to have Iraqi forces lead a campaign with our forces in support to protect the population of Baghdad from intimidation and violence instigated by Sunni and Shia extremist groups and to enable the Iraqi government to take the difficult steps necessary to address that nation's underlying issues.

This means, above all, strengthening those in Iraq who are prepared to address its problems peacefully against those who seek only violence, death and chaos.

The term "surge" has been used in relation to increasing U.S. troop levels, and an increase certainly will take place. But what is really going to take place, and what is going to take place, is a "surge" across all lines of operations—military and non-military, Iraq and Coalition. The President's plan has Iraqis in the lead and seeks a better balance of U.S. military and non-military efforts than was the case in the past. We cannot succeed in Iraq without the important non-military elements Secretary Rice just mentioned.

The increase in military forces will be phased in. It will not unfold overnight. There will be no "D-Day." It won't look like the Gulf War.

The timetable for the introduction of the additional U.S. forces will provide ample opportunity early on—and before many of the additional U.S. troops arrive in Iraq—to evaluate the progress of this endeavor and whether the Iraqis are fulfilling their commitments to us.

This updated plan builds on the lessons and experiences of the past. It places new emphasis on and adds new resources to the "holding" and "building" part of the "clear, hold and build" strategy.

At this pivotal moment, the credibility of the United States is on the line in Iraq. Governments in the region—both friends and adversaries—are watching what we do and will draw their own conclusions about our resolve and the steadfastness of our commitments.

Whatever one's views on how we got to this point in Iraq, there is widespread agreement that failure there would be a calamity that would haunt our nation in the region. The violence in Iraq, if unchecked, could spread outside its borders and draw other states into a regional conflagration. In addition, one would see:

- An emboldened and strengthened Iran;
- A safehaven and base of operations for jihadist networks in the heart of the Middle East;
- A humiliating defeat in the overall campaign against violent extremism worldwide; and
- An undermining of the credibility of the United States.

Given what is at stake, failure in Iraq is not an option. I would like to conclude my remarks with two announcements.

First, the President announced last night that he would strengthen our military for the long war against terrorism by authorizing an increase in the overall strength of the Army and the Marine Corps. I am recommending to him a total increase in the two services of 92,000 soldiers and Marines over the next five years—65,000 soldiers and 27,000 Marines. The emphasis will be on increasing combat capability.

This increase will be accomplished in two ways. First, we will propose to make permanent the temporary increase of 30,000 for the Army and 5,000 for the Marine Corps. Then we propose to build up from that base in annual increments of 7,000 troops a year for the Army, and 5,000 for the Marine Corps until they reach a level of 202,000. And the Army would be at 507,000.

We should recognize that while it may take some time for these new troops to become available for deployment, it is important that our men and women in uniform know that additional manpower and resources are on the way.

Second, for several months, the Department of Defense has been assessing whether we have the right policies to govern how we manage and deploy members of the Reserves, the National Guard and our active component units.

Based on this assessment and the recommendations of our military leadership, I am making the following changes in Department policy.

First, the mobilization of ground reserve forces going forward will be managed on a unit instead of an individual basis. This change will allow us to achieve greater unit cohesion and predictability in how reserve units train and deploy.

Second, from this point forward, members of the Reserves will be involuntarily mobilized for a maximum of one year at a time, in contrast to the current practice of 16 to 24 months.

Third, the planning objective for guard and reserve units will remain one year of being mobilized followed by five years demobilized. However, today's global demands will require a number of selected guard and reserve units to be remobilized sooner than this standard. Our intention is that such exceptions be temporary. The goal for the active force rotation cycle remains one year deployed for every two years at home station. Today, most active units are receiving only one year at home station before deploying again. Mobilizing select guard and reserve units before this five year period is complete will allow us to move closer to relieving the stress on the total force.

Fourth, I am directing the establishment of a new program to compensate individuals in both the active and reserve components who are required to mobilize or deploy early or extend beyond the established rotation policy goals.

Fifth, I am also directing that all commands and units review how they administer the hardship waiver program to ensure that they are properly taking into account exceptional circumstances facing military families of deployed service members.

It is important to note that these policy changes have been under discussion for some time within the Department of Defense and would be needed independently of the President's announcement on Iraq last night.

And there will be a handout on the details as they are complicated.

Finally, I am also pleased to report that all active branches of the U.S. military exceeded their recruiting goals for the month of December, with particularly strong showings by the Army and the Marine Corps. Our nation is truly blessed that so many talented and patriotic young people have stepped forward to defend our nation, and that so many servicemen and women have chosen to continue to serve.

Thank you, and we will be happy to take your questions.

## Testimony on Iraq to the House Armed Services Committee

*Washington, DC, Thursday, January 11, 2007*

Thank you.

Mr. Chairman, Senator McCain, members of the committee.

Let me say at the outset that it's a pleasure to appear before this committee for the first time as secretary of Defense. The Senate Armed Services Committee has long been a steadfast friend and ally of our men and women in uniform, and a source of steadfast support in meeting the nation's defense needs, and I thank you for that and I look forward to working with you.

Let me begin by quickly summarizing two announcements I made yesterday morning.

One of which Senator McCain just referred to. I have recommended to the President an increase in the two services of 92,000 soldiers and Marines over the next five years—65,000 soldiers and 27,000 Marines. The emphasis will be on increasing combat capability.

The increase will be accomplished in two ways. First, we'll make permanent the temporary increase of 30,000 for the Army and 5,000 for the Marine Corps. We then propose to build up from that base in annual increments over five years 7,000 troops a year for the Army until they reach 547,000 and 5,000 a year for the Marine Corps until they reach 202,000.

While it may take some time for these troops to become available for deployment, it is important for our men and women in uniform to know that additional manpower and resources are on the way.

Second, for several months the Department has been assessing whether we have the right policies to govern how we manage and deploy members of the Reserves, the National Guard and our active component units.

Based on this assessment and the recommendations of our military leadership, I am prepared to make the following changes to Department policy:

First, mobilization of the ground Reserve forces will be managed on a unit basis instead of an individual basis. This change will allow us to achieve greater unit cohesion and predictability in how Reserve units train and deploy.

Second, from this point forward, members of Reserves will be involuntarily mobilized for a maximum of one year at any one time, in contrast to the current practice of 16 to 24 months.

Third, the planning objective for Guard and Reserve units will remain one year of being mobilized followed by five years demobilized. However, today's global demands will require a number of selected Guard and Reserve units to be remobilized sooner than this standard. Our intention is that such exceptions will be temporary.

The goal for the active force rotation cycle remains one year deployed for every two years at home station. Today, most active units are receiving one year at home station before deploying again. We believe that mobilizing select Guard and Reserve units before their five-year period is complete will allow us to move closer to relieving the stress on the active—on the total force.

Fourth, I'm directing the establishment of a new program to compensate individuals in both the active and Reserve components that are required to mobilize or deploy early or extend beyond the established rotation policy goals.

Fifth and finally, I'm directing that all commands and units review how they administer the hardship waiver program to ensure that they are properly taking into account exceptional circumstances facing military families of deployed service members.

It's important to note that these policy changes have been under discussion for some time within the Department of Defense, and would need to take place regardless of the President's announcement the other night.

Just as an aside, but an important one, I'm pleased to report to the committee that all active branches of the military exceeded their recruiting goals for the month of December, with particularly strong showings by the Army and the Marine Corps. Our nation is truly blessed that so many talented and patriotic young people have stepped forward to defend our nation and that so many servicemen and women have chosen to continue to serve.

With respect to the President's initiative, he described a new way forward in Iraq on Wednesday night, a new approach to overcoming the steep challenges facing us in that country and that part of the world.

I know many of you have concerns about the new strategy in Iraq and in particular are skeptical of the Iraqi government's will and ability to act decisively against sectarian violence and are skeptical as well about a commitment of additional troops.

The President and his national security team have had the same concerns as we have debated and examined our options going forward. And yet our commanders on the ground, and the President's intended nominee as the new commander, believe this is a sound plan, in no small part because General Casey and other senior military officers have worked closely with the Iraqi government in developing it.

Further, the President, Ambassador Khalilzad, and General Casey have had prolonged and extremely candid conversations, not just with Prime Minister Maliki, but with other senior leaders of the Iraqi government, and have come away persuaded that they finally have the will to act against all instigators of violence in Baghdad.

This is, I think, a pivot point—the pivot point—in Iraq, as the Iraqi government insists on assuming the mantle of leadership in the effort to regain control of its own capital. I want you do know that the time table for the introduction of additional U.S. forces will provide ample opportunity early on and before many of the additional U.S. troops arrive in Iraq—to evaluate the progress of this endeavor and whether the Iraqis are fulfilling their commitments to us.

Let me make two other quick points. First, this strategy entails a strengthening across all aspects of the war effort—military and non-military—including the economic, governance and political areas. Overcoming the challenges in Iraq cannot be achieved simply by military means—no matter how large or sustained—without progress by the Iraqis in addressing the underlying issues dividing the country.

Second, we must keep in mind the consequences of an American failure in Iraq. As I said in my confirmation hearing, developments in Iraq over the next year or two will shape the future of the Middle East and impact global geopolitics for a long time to come.

I would not have taken this position if I did not believe that the outcome in Iraq will have a profound and long-lived impact on our national interests.

Mistakes certainly have been made by the United States in Iraq, but however we got to this moment, the stakes now are incalculable.

Your senior professional military officers in Iraq and in Washington believe in the efficacy of the strategy outlined by the President. They believe it is a sound plan that can work if the Iraqi government follows through on its commitments and if the non-military aspects of the strategy are implemented and sustained.

Our senior military officers have worked closely with the Iraqis to develop this plan. The impetus to add U.S. forces came initially from our commanders there. It would be a sublime yet historic irony if those who believe the views of the military professionals were neglected at the onset of the war were now to dismiss the views of the military as irrelevant or wrong.

Thank you, Mr. Chairman.

## Statement on Iraq to the Senate Armed Service Committee

*Washington, DC, Friday, January 12, 2007*

THANK you.

Mr. Chairman, Senator McCain, members of the committee.

Let me say at the outset that it's a pleasure to appear before this committee for the first time as secretary of Defense. The Senate Armed Services Committee has long been a steadfast friend and ally of our men and women in uniform, and a source of steadfast support in meeting the nation's defense needs, and I thank you for that and I look forward to working with you.

Let me begin by quickly summarizing two announcements I made yesterday morning.

One of which Senator McCain just referred to. I have recommended to the President an increase in the two services of 92,000 soldiers and Marines over the next five years—65,000 soldiers and 27,000 Marines. The emphasis will be on increasing combat capability.

The increase will be accomplished in two ways. First, we'll make permanent the temporary increase of 30,000 for the Army and 5,000 for the Marine Corps. We then propose to build up from that base in annual increments over five years 7,000 troops a year for the Army until they reach 547,000 and 5,000 a year for the Marine Corps until they reach 202,000.

While it may take some time for these troops to become available for deployment, it is important for our men and women in uniform to know that additional manpower and resources are on the way.

Second, for several months the Department has been assessing whether we have the right policies to govern how we manage and deploy members of the Reserves, the National Guard and our active component units.

Based on this assessment and the recommendations of our military leadership, I am prepared to make the following changes to Department policy:

First, mobilization of the ground Reserve forces will be managed on a unit basis instead of an individual basis. This change will allow us to achieve greater unit cohesion and predictability in how Reserve units train and deploy.

Second, from this point forward, members of Reserves will be involuntarily mobilized for a maximum of one year at any one time, in contrast to the current practice of 16 to 24 months.

Third, the planning objective for Guard and Reserve units will remain one year of being mobilized followed by five years demobilized. However, today's global demands will require a number of selected Guard and Reserve units to be remobilized sooner than this standard. Our intention is that such exceptions will be temporary.

The goal for the active force rotation cycle remains one year deployed for every two years at home station. Today, most active units are receiving one year at home station before deploying again. We believe that mobilizing select Guard and Reserve units before their five-year period is complete will allow us to move closer to relieving the stress on the active—on the total force.

Fourth, I'm directing the establishment of a new program to compensate individuals in both the active and Reserve components that are required to mobilize or deploy early or extend beyond the established rotation policy goals.

Fifth and finally, I'm directing that all commands and units review how they administer the hardship waiver program to ensure that they are properly taking into account exceptional circumstances facing military families of deployed service members.

It's important to note that these policy changes have been under discussion for some time within the Department of Defense, and would need to take place regardless of the President's announcement the other night.

Just as an aside, but an important one, I'm pleased to report to the committee that all active branches of the military exceeded their recruiting goals for the month of December, with particularly strong showings by the Army and the Marine Corps. Our nation is truly blessed that so many talented and patriotic young people have stepped forward to defend our nation and that so many servicemen and women have chosen to continue to serve.

With respect to the President's initiative, he described a new way forward in Iraq on Wednesday night, a new approach to overcoming the steep challenges facing us in that country and that part of the world.

I know many of you have concerns about the new strategy in Iraq and in particular are skeptical of the Iraqi government's will and ability to act decisively against sectarian violence and are skeptical as well about a commitment of additional troops.

The President and his national security team have had the same concerns as we have debated and examined our options going forward. And yet our commanders on the ground, and the President's intended nominee as the new commander, believe this is a sound plan, in no small part because General Casey and other senior military officers have worked closely with the Iraqi government in developing it.

Further, the President, Ambassador Khalilzad, and General Casey have had prolonged and extremely candid conversations, not just with Prime Minister Maliki, but with other senior leaders of the Iraqi government, and have come away persuaded that they finally have the will to act against all instigators of violence in Baghdad.

This is, I think, a pivot point—the pivot point—in Iraq, as the Iraqi government insists on assuming the mantle of leadership in the effort to regain control of its own capital. I want you do know that the time table for the introduction of additional U.S. forces will provide ample opportunity early on and before many of the additional U.S. troops arrive in Iraq—to evaluate the progress of this endeavor and whether the Iraqis are fulfilling their commitments to us.

Let me make two other quick points. First, this strategy entails a strengthening across all aspects of the war effort—military and non-military—including the economic, governance and political areas. Overcoming the challenges in Iraq cannot be achieved simply by military means—no matter how large or sustained—without progress by the Iraqis in addressing the underlying issues dividing the country.

Second, we must keep in mind the consequences of an American failure in Iraq. As I said in my confirmation hearing, developments in Iraq over the next year or two will shape the future of the Middle East and impact global geopolitics for a long time to come.

I would not have taken this position if I did not believe that the outcome in Iraq will have a profound and long-lived impact on our national interests.

Mistakes certainly have been made by the United States in Iraq, but however we got to this moment, the stakes now are incalculable.

Your senior professional military officers in Iraq and in Washington believe in the efficacy of the strategy outlined by the President. They believe it is a sound plan that can work if the Iraqi government follows through on its commitments and if the non-military aspects of the strategy are implemented and sustained.

Our senior military officers have worked closely with the Iraqis to develop this plan. The impetus to add U.S. forces came initially from our commanders there. It would be a sublime yet historic irony if those who believe the views of the military professionals were neglected at the onset of the war were now to dismiss the views of the military as irrelevant or wrong.

Thank you, Mr. Chairman.

# Posture Statement to the Senate Armed Services Committee

*Washington, DC, Tuesday, February 06, 2007*

Mr. Chairman, Senator McCain, Members of the Committee.

I thank the Committee for all you have done to support our military these many years, and I appreciate the opportunity to provide an overview of the way ahead at the Department of Defense through the budgets being proposed this week: First, the President's Fiscal Year 2008 Defense Budget, which includes the base budget request and the FY 2008 Global War on Terror Request; and second, the FY 2007 Emergency Supplemental Appropriation Request to fund war-related costs for the remainder of this fiscal year.

I believe it is important to consider these budget requests in some historical context as there has been, understandably, some element of sticker shock at their combined price tags—more than $700 billion in total.

But, consider that at about 4 percent of America's Gross Domestic Product, the amount of money the United States is expected to spend on defense this year is actually a smaller percentage of GDP than when I left government 14 years ago following the end of the Cold War—and a significantly smaller percentage than during previous times of war, such as Vietnam and Korea.

Since 1993, with a defense budget that is a smaller relative share of our national wealth, the world has gotten more complicated, and arguably more dangerous. In addition to fighting the Global War on Terror, we also face:

- The danger posed by Iran and North Korea's nuclear ambitions, and the threat they pose not only to their neighbors, but globally, because of their record of proliferation.
- The uncertain paths of China and Russia, which are both pursuing sophisticated military modernization programs; and
- A range of other potential flashpoints and challenges.

In this strategic environment, the resources we devote to defense should be at the level to adequately meet those challenges.

Someone once said that "Experience is that marvelous thing that enables you to recognize a mistake when you make it again."

Five times over the past 90 years the United States has either slashed defense spending or disarmed outright in the mistaken belief that the nature of man or behavior of nations had somehow changed, or that we would no longer need capable, well funded military forces on hand to confront threats to our nation's interests and security. Each time we have paid a price.

The costs of defending our nation are high. The only thing costlier, ultimately, would be to fail to commit the resources necessary to defend our interests around the world, and to fail to prepare for the inevitable threats of the future.

## FY 2008 Base Budget

The President's FY 2008 base budget request of $481.4 billion is an increase of 11.3 percent over the projected enacted level of FY 2007, and provides the resources needed to man, organize, train, and equip the Armed Forces of the United States. This budget continues efforts to reform and transform our military establishment to be more agile, adaptive, and expeditionary to deal with a range of both conventional and irregular threats.

Some military leaders have argued that while our forces can support current operations in the War on Terror, these operations are increasing risks associated with being called on to undertake a major conventional conflict elsewhere around the world. This budget provides additional resources to mitigate those risks.

The FY 2008 base budget includes increases of about $16.8 billion over last year for investments in additional training, equipment repair and replacement, and intelligence and support. It provides increases in combat training rotations, sustains air crew training, and increases ship steaming days.

## Increase Ground Forces

Despite significant improvements in the way our military is organized and operated, the ongoing conflicts in Iraq and Afghanistan have put stress on parts of our nation's ground forces.

Last month, the President called for an increase in the permanent active end strength of the Army and Marine Corps of some 92,000 troops by FY 2012. The base budget request adds $12.1 billion to increase ground forces in the next fiscal year, which will consist of 7,000 additional Soldiers and 5,000 additional Marines.

Special Operations Forces, who have come to play an essential and unique role in operations against terrorist networks, will also grow by 5,575 troops between FY 2007 and FY 2008.

## Strategic Investments—Modernization

The base budget invests $177 billion in procurement and research and development that includes major investments in the next generation of technologies. The major weapons systems include:
- Future Combat System ($3.7 billion)—The first comprehensive modernization program for the Army in a generation.
- Joint Strike Fighter ($6.1 billion)—This next generation strike aircraft has variants for the Air Force, the Navy, and the Marine Corps. Eight international partners are contributing to the JSF's development and production.
- F-22A ($4.6 billion)—Twenty additional aircraft will be procured in FY 2008.
- Shipbuilding ($14.4 billion)—The increase of $3.2 billion over last year is primarily for the next generation aircraft carrier, the CVN-21, and the LPD-17 amphibious transport ship. The long-term goal is a 313-ship Navy by 2020.

## Missile Defense

I have believed since the Reagan administration that if we can develop a missile defense capability, it would be a mistake for us not to do so. There are many countries that either have or are developing ballistic missiles, and there are at least two or three others—including North Korea—that are developing longer-range systems. We also have an obligation to our allies, some of whom have signed on as partners in this effort. The Department is proceeding with formal negotiations with Poland and the Czech Republic on establishing a European missile site. The missile defense program funded by this request will continue to test our capability against more complex and realistic scenarios. I urge the committee to approve the full $9.9 billion requested for the missile defense and Patriot missile programs.

## Space Capabilities

The recent test of an anti-satellite weapon by China underscored the need to continue to develop capabilities in space. The policy of the U.S. Government in this area remains consistent with the long-standing principles that were established during the Eisenhower Administration, such as the right of free passage and the use of space for peaceful purposes. Space programs are essential to the U.S. military's communications, surveillance, and reconnaissance capabilities. The base budget requests about $6.0 billion to continue the development and fielding of systems that will maintain U.S. supremacy while ensuring unfettered, reliable, and secure access to space.

## Recapitalization

A major challenge facing our military is that several key capabilities are aging and long overdue for being replaced. The prime example is the Air Force KC-135 tanker fleet, which averages 45 years per plane. It is becoming more expensive to maintain and less reliable to operate. The Air Force has resumed a transparent and competitive replacement program to recapitalize this fleet with the KC-X aircraft. The KC-X will be able to carry cargo and passengers and will be equipped with defensive systems. It is the U.S. Transportation Command's and the Air Force's top acquisition and recapitalization priority.

## Quality of Life—Sustaining the All-Volunteer Force

Our nation is fortunate that so many talented and patriotic young people have stepped forward to serve, and that so many of them have chosen to continue to serve.

In December, all active branches of the U.S. military exceeded their recruiting goals, with particularly strong showings by the Army and Marine Corps. The FY 2008 request includes $4.0 billion for recruiting and retention to ensure that the military continues to attract and retain the people we need to grow the ground forces and defend the interests of the United States.

We will continue to support the all-volunteer force and their families through a variety of programs and initiatives. The budget includes:

- $38.7 billion for health care for both active and retired service members;
- $15 billion for Basic Allowance for Housing to ensure that, on average, troops are not forced to incur out-of-pocket costs to pay for housing;
- $2.9 billion to improve barracks and family housing and privatize an additional 2,870 new family units; and
- $2.1 billion for a 3 percent pay increase for military members.

In addition, recently announced changes in the way the military uses and employs the Reserves and National Guard should allow for a less frequent and more predictable mobilization schedule for our citizen soldiers.

Combined with other initiatives to better organize, manage, and take care of the force, these recent changes should mean that in the future our troops should be deployed or mobilized less often, for shorter periods of time, and with more predictability and a better quality of life for themselves and their families.

## Train and Equip Authorities

Operations in Iraq, Afghanistan, and elsewhere have shown the critical importance of building the capacity and capability of partners and allies to better secure and govern their own countries.

In recent years we have struggled to overcome the patchwork of authorities and regulations that were put in place during a very different era—the Cold War—to confront a notably different set of threats and challenges.

The administration has, with congressional support, taken some innovative steps to overcome these impediments. A significant breakthrough was the Section 1206 authority that allows the Defense and State Departments to more rapidly and effectively train and equip partner military forces. In the FY 2008 base budget, we are seeking dedicated funding of $500 million to use this authority. I would ask for a serious, collaborative effort with Congress to develop the right interagency funding mechanisms and authorities to meet critical national security needs.

## Global War on Terror Requests

The President's two war-related requests are the FY 2007 Emergency Supplemental Request for $93.4 billion, and the FY 2008 Global War on Terror Request for $141.7 billion.

The FY 2007 Supplemental Request is in addition to the $70 billion that has already been appropriated for war-related costs in this fiscal year. If these additional funds are delayed, the military will be forced to engage in costly and counterproductive reprogramming actions starting this spring to make up the shortfall. Timely enaction of the FY 2007 Supplemental is critical to ensuring our troops in the field have the resources they need.

The additional U.S. ground and naval forces being sent to the Iraq theater are projected to cost $5.6 billion. This total includes funding for personnel costs, supplies, spare parts, contractor support, and transportation.

The FY 2008 GWOT Request complies with Congress's direction to include the costs of ongoing operations in Iraq and Afghanistan in the annual Defense Department budget. Given the uncertainty of projecting the cost of operations so far in the future, the funds sought for the FY 2008 GWOT Request are generally based on a straight-line projection of current costs for Iraq and Afghanistan.

The war-related requests include $39.3 billion in the FY 2007 Supplemental and $70.6 billion in the FY 2008 GWOT Request to provide the incremental pay, supplies, transportation, maintenance and logistical support to conduct military operations.

## Reconstitution

The FY 2007 Supplemental requests $13.9 billion—and the FY 2008 GWOT Request $37.6 billion—to reconstitute our nation's armed forces—in particular, to refit the ground forces, the Army and Marine Corps, who have borne the brunt of combat in both human and material terms. These funds will go to repair or replace equipment that has been destroyed, damaged, or stressed in the current conflict.

All Army units deployed, or about to deploy, for missions overseas are fully trained and equipped, often with additional gear for their particular mission. In an expeditionary, rotational force one can expect that units returning from their deployment will decline to a lower readiness level as personnel turn over and equipment is repaired or replaced. The $13.6 billion in reset funds in the FY 2008 GWOT Request for the U.S. Army will go a long way towards raising the readiness levels across the force.

## Force Protection

The war-related requests include $10.4 billion in the FY 2007 Supplemental, and $15.2 billion in the FY 2008 GWOT Request for investments in new technologies to better protect our troops from an agile and adaptive enemy. Programs being funded would include a new generation of body armor, vehicles that can better withstand the blasts from Improved Explosive Devises (IEDs), and electronic devices that interrupt the enemy's ability to attack U.S. forces. Within this force protection category, the FY 2007 Supplemental includes $2.4 billion and the FY 2008 GWOT includes $4.0 billion to counter and defeat the threat posed by IEDs.

## Afghan/Iraqi Security Forces

The FY 2007 Supplemental requests $9.7 billion, and the FY 2008 GWOT requests $4.7 billion, to stand up capable military and police forces in Afghanistan and Iraq.

The bulk of these funds are going to train and equip Afghan National Security Forces (ANSF) to assume the lead in operations throughout Afghanistan. As of last month, some 88,000 have been trained and equipped, an increase of 31,000 from the previous year.

The $5.9 billion for the ANSF in the FY 2007 Supplemental is a substantial increase over previous years' appropriations. It reflects the urgent priority of countering increased activity by the Taliban, Al Qaeda, and narcotics traffickers to destabilize and undermine the new democracy in Afghanistan. These funds will significantly upgrade the capability of Afghan forces to conduct independent counter-insurgency operations.

In Iraq, more than 300,000 soldiers and police have been trained and equipped, and are in charge of more than 60 percent of Iraqi territory and more than 65 percent of that country's population. They have assumed full security responsibility for three out of Iraq's 18 provinces and are scheduled to take over more territory over the course of the year. These Iraqi troops, though far from perfect, have shown that they can perform with distinction when properly led and supported. Iraqi forces will be in the lead during operations to secure Baghdad's violent neighborhoods. By significantly increasing and improving the embedding program, Iraqi forces will operate with more and better Coalition support than they had in the past.

## Non-Military Assistance

Success in the kinds of conflicts our military finds itself in today—in Iraq, or elsewhere—cannot be achieved by military means alone. The President's strategy for Iraq hinges on key programs and additional resources to improve local governance, delivery of public services, and quality of life—to get angry young men off the street and into jobs where they will be less susceptible to the appeals of insurgents or militia groups.

Commanders Emergency Response Program, or (CERP) funds are a relatively small piece of the war-related budgets—$456 million in the FY 2007 Supplemental, and $977 million in the FY 2008 GWOT Request. But because they can be dispensed quickly and applied directly to local needs, they have had a tremendous impact—far beyond the dollar value—on the ability of our troops to succeed in Iraq and Afghanistan. By building trust and confidence in Coalition forces, these CERP projects increase the flow of intelligence to commanders in the field and help turn local Iraqis and Afghans against insurgents and terrorists.

## Conclusion

With the assistance and the counsel of Congress, I believe we have the opportunity to do right by our troops and the sacrifices that they and their families have made these past few years. That means we must make the difficult choices and commit the necessary resources to not only prevail in the current conflicts in which they are engaged, but to be prepared to take on the threats that they, their children, and our nation may face in the future.

# Munich Conference on Security Policy

*Munich, Germany, Sunday, February 11, 2007*

THANK you, Horst.

Distinguished ministers, parliamentarians, representatives of the United States Congress—ladies and gentlemen.

I would like to thank Horst for inviting me to speak at this venerable forum to offer some thoughts on our transatlantic partnership. It's gratifying to see so many people who I've worked with on these security issues going back many years. Speaking of issues going back many years, as an old Cold Warrior, one of yesterday's speeches almost filled me with nostalgia for a less complex time. Almost.

Many of you have backgrounds in diplomacy or politics. I have, like your second speaker yesterday, a starkly different background—a career in the spy business. And, I guess, old spies have a habit of blunt speaking.

However, I have been to re-education camp, spending four and half years as a university president and dealing with faculty. And, as more than a few university presidents have learned in recent years, when it comes to faculty it is either "be nice" or "be gone."

The real world we inhabit is a different and a much more complex world than that of 20 or 30 years ago. We all face many common problems and challenges that must be addressed in partnership with other countries, including Russia.

For this reason, I have this week accepted the invitation of both President Putin and Minister of Defense Ivanov to visit Russia. One Cold War was quite enough.

The world has dramatically changed since May 1989, when Horst Teltschik and I sat out on the patio of the "Chancellor's Bungalow" in Bonn with Chancellor Kohl and my colleague Larry Eagleburger. At that time, the allies were trying to come together on the issue of reducing conventional forces in Europe. The way I remember that particular meeting, however, was that the tough part wasn't addressing the military balance of power in Europe, it was seeing to it that there were enough cakes and pastries on hand for both the Chancellor and the Deputy Secretary of State.

It is certainly good to be in Munich following the NATO ministerial in Seville. I should say that this trip has been quite a different experience from my so-called fact-finding excursion last month to Europe, the Middle East, and Central Asia. The one fact, above all, that became clear from that venture is that I am too old to visit seven countries in five days. However, I have now learned here in Munich that I am still too young to sit still for seven hours.

As many of you know, the security of this continent has been of interest to me for much of my academic and professional life—for more than 40 years in fact. This was true when I was a Ph.D. candidate in Russian and Soviet history, through my career at CIA, as well as during service on the National Security Council under four presidents.

For many of those years, I worked hand in hand with colleagues from Western European governments to help coordinate our actions and responses in the latter half of the Cold War. Many of those colleagues are here this morning.

I had a ringside seat for an extraordinary run of events from the 1975 Helsinki conference to the liberation of Central and Eastern Europe a decade and a half later.

During that struggle, there were times of confrontation between the superpowers. Relations among the allies were not without their stresses and strains, either. But our Atlantic partnership was strong enough to allow us to surmount the difficulties and make the right choices at the right times. For example, the decision to deploy cruise and Pershing missiles to counter the Soviet Union's new weapons in the late 1970s, was politically difficult for many allies. But ultimately, the courage and leadership of statesmen and stateswomen on both sides of the Atlantic, and the actual deployment of the missiles early in the 1980s, helped set the stage for deep reductions in nuclear arms and the end of the Cold War.

Looking back, it seems clear that totalitarianism was defeated as much by ideas the West championed—then as now—as by ICBMs, tanks, and warships that the West deployed. Our most effective weapon, then and now, has been Europe's and North America's shared belief in political and economic freedom, religious toleration, human rights, representative government, and the rule of law. These values kept our side united, and inspired those on the other side—in Wenceslas Square, in Gdansk, behind the wall in Berlin, and in so many other places around the world—to defeat communism from within.

At the end, the peoples of Eastern Europe and the former Soviet Union simply stood up, shrugged off their chains, and reclaimed a future based on these same ideas.

I believe these shared values and shared interests endure, as do our shared responsibilities to come to their defense. Today, they are under threat by another virulent ideological adversary and are confronted by a range of other looming geopolitical challenges.

This strategic environment has challenged the mission and identity of the Atlantic Alliance—an institution and an arrangement that, in my view, is the political and military expression of a deeper bond between Europe and North America.

Many of these questions are not new. I recall spending countless hours beginning in 1989 on the future of the Alliance and how it would need to change in order to remain vital and relevant after the collapse of the Warsaw Pact.

The question that still confronts us today is how a partnership originally formed to defend fixed borders should adapt to an era of unconventional and global threats. The European continent, of course, has been confronting the threat of terrorism for decades. I don't have to remind the citizens of Munich of this—the very city where, in 1972, the world witnessed the kidnapping and massacre of Olympic athletes not too far from where we sit today.

But the challenge posed by violent extremism today is unlike anything the West has faced in many generations. In many ways it is grounded in a profound alienation from the foundations of the modern world—religious toleration, freedom of expression and equality for women. As we have seen, many of these extremist networks are homegrown, and can take root in the restless and alienated immigrant populations of Europe.

The dark talent of the extremists today is, as President Bush said, to combine "new technologies and old hatreds." Their ability to tap into global communications systems turns modern advances against us and turns local conflicts into problems potentially of much wider concern. The interest they have shown in weapons of mass destruction is real and needs to be taken seriously.

We have learned that from a distant and isolated place, from any failed or extremist state—such as Afghanistan during the 1990s—these networks can plan and launch far-reaching and devastating attacks on free and civilized nations.

No fewer than 18 terrorist organizations, many linked with al Qaeda, have pulled off bloody attacks throughout the world—in the United States, Spain, the United Kingdom, India, Algeria, Somalia, Russia, Pakistan, Jordan, Egypt, Indonesia, Tunisia, Morocco and in others as well.

Those attacks—and other threats that have since emerged—revealed even more starkly the need to reorient the Atlantic Alliance to be able to export security beyond the borders of NATO.

Although NATO was created to oppose Soviet communism, its guiding principle was a broad and deep one from the very start: to build a defensive alliance against any threat to the security and interests of the transatlantic community for generations to come.

And today we see that an Alliance that never fired a shot in the Cold War now conducts six missions on three continents. It has created new mechanisms for action on the international stage. It has been through profound changes and will undergo more in the future.

We see this in NATO's truly historic mission in Afghanistan, where Alliance forces have engaged in significant ground combat for the first time, in complex operations across difficult terrain, in a theater many long miles from Western Europe.

Last year in Afghanistan, the Taliban paid the price for testing the fighting mettle of NATO forces, as troops from the United Kingdom, Canada, the Netherlands, Australia, Romania, Estonia and Denmark—along with our Afghan allies—prevailed in often fierce combat in Kandahar province.

In fact, as the NATO allies just discussed in Seville, if we take the necessary steps now, the offensive in Afghanistan this spring will be our offensive—one that will inflict a powerful setback on the enemy of an elected government supported by the overwhelming majority of the Afghan people.

Going forward, it is vitally important that the success Afghanistan has achieved not be allowed to slip away through neglect or lack of political will or resolve.

All allies agree we need a comprehensive strategy—combining a muscular military effort with effective support for governance, economic development, and counter-narcotics.

But now we have to back up those promises with money and with forces. An Alliance consisting of the world's most prosperous industrialized nations, with over two million people in uniform—not even counting the American military—should be able to generate the manpower and materiel needed to get the job done in Afghanistan—a mission in which there is virtually no dispute over its justness, necessity, or international legitimacy. Our failure to do so would be a mark of shame.

What has emerged in Afghanistan is a test of our ability to overcome a challenge of enormous consequence to our shared values and interests. In today's strategic environment, there are potentially others:

The fault lines of sectarian conflict and jihadist movements radiating outward from the Middle East and Central Asia; an Iran with hegemonic ambitions seeking nuclear weapons; and the struggle over the future of Iraq, with enormous implications for our common interests in the Middle East—and beyond.

Looking eastward, China is a country at a strategic crossroads. All of us seek a constructive relationship with China, but we also wonder about the strategic choices China may make. We note with concern its recent test of an anti-satellite weapon.

Russia is a partner in endeavors. But we wonder, too, about some Russian policies that seem to work against international stability, such as its arms transfers and its temptation to use energy resources for political coercion. And as the NATO Secretary General said yesterday, Russia need not fear law-based democracies on its borders.

In this strategic environment, the Alliance must be willing to alter long-standing habits, assumptions and arrangements. Much progress has been made, to be sure. After almost 15 years away from government, I have been deeply impressed by the new expeditionary capabilities and institutional reforms NATO has undertaken. The missile defense discussion the United States is having with Poland, the Czech Republic, the U.K, and Denmark to protect our homelands is another promising development. And, at the Riga Summit, our allied leaders agreed to strengthen our security relationships with like-minded nations in other parts of the globe—such as Australia, Japan, and South Korea.

But in addition to pursuing new missions, capabilities and partnerships, the members of this Alliance must, individually and collectively, be willing to commit the necessary resources as well—not just in Afghanistan, but across the board.

The benchmark of spending 2 percent of Gross Domestic Product on defense, for example, is a commitment agreed to by each member. Such an investment is necessary to meet our collective obligations to ensure that when we stand together in battle—whether in Afghanistan or elsewhere—the quality, quantity and sophistication of our equipment and our capabilities are at an appropriate level. And yet, at this time, only six of NATO's 26 members have met the GDP standard.

Over the years, people have tried to put the nations of Europe and the Alliance into different categories:

The "free world" versus "those behind the Iron Curtain";

"North" versus "South";

"East" versus "West";

I am told that some have even spoken in terms of "old" Europe versus "new."

All of these characterizations belong to the past. The distinction I would draw is a very practical one—a "realist's" view: It is between Alliance members who do all they can to fulfill collective commitments, and those who do not. NATO is not a "paper membership," or a "social club," or a "talk shop." It is a military alliance—one with very serious real-world obligations.

It is a sad reality today, as through all human history, that some seek through violence and crimes against the innocent to dominate others. Another sad reality is that, when all is said and done, they understand and bow not to reason nor to negotiation, but only to superior force. This is perhaps politically incorrect, and perhaps an old intelligence officer being too blunt. But it is reality.

And it is the power, the political and military power of 26 democracies of NATO—the most potent Alliance in the history of the world—that is the shield behind which the ideas and values we share are spreading around the globe.

In short, meeting our commitment to one another and to those we strive to help—from the Balkans to Afghanistan and beyond—is critical to our success and theirs.

Looking back, the Cold War was an epic struggle that incurred epic costs. I believe we all agree that incurring those costs was preferable to the alternatives: catastrophic conflict or totalitarian domination.

The range of challenges and threats we face today will also test our willingness to meet our commitments to spend the money and take the risks—indeed, to fully embrace our shared responsibility to protect our shared interests and values.

There cannot be any doubt: The world needs a vibrant and muscular transatlantic alliance. The cooperation between our countries must continue and it must deepen. We will need to work hard at it. And we are working hard together in the Balkans, in Afghanistan, and, many of us, in Iraq.

As we face these challenges as rich and powerful democracies, it is worth recalling the words of a leader of a fledgling and weak alliance of disparate provinces with:

Disrupted economies; differing issues and goals; diverse allegiances; mutual suspicion; an army comprised of soldiers often with parochial loyalties, and lacking in equipment and training; and with but one strong ally.

George Washington reminded his countrymen—and us—that "Perseverance and spirit have done wonders in all ages." These should be our watchwords going forward: "Perseverance" and "spirit." And, I should add—"unity."

Thank you very much. I look forward to your questions.

## U.S. Senate Committee on Appropriations

*Washington, DC, Tuesday, February 27, 2007*

MR. Chairman, Senator Cochran, Members of the Committee:

I appreciate the opportunity to join Secretary Rice in discussing the President's Supplemental Appropriation Request to fund the costs of operations in Iraq, Afghanistan and the wider Global War on Terror.

From the start, I would like to express my strong support for the programs funded in the State Department's request. The kinds of challenges our country faces in Iraq and Afghanistan cannot be overcome without the important non-military efforts outlined by Secretary Rice.

The 2007 Supplemental Request of $93.4 billion for the Department of Defense is in addition to the $70 billion that has already been appropriated for war-related costs in this fiscal year. If these additional funds are delayed, the military will be forced to engage in costly and counterproductive re-programming actions starting this spring to make up the shortfall. Timely enactment of this Supplemental Request is critical to ensuring our troops in the field have the resources they need.

While our country is properly focused on the serious situation in Iraq, it is critical that the gains made in Afghanistan these past few years not be allowed to slip away. This was at the top of my agenda at the NATO ministerial earlier this month in Seville.

I believe that it is important to consider the defense budget requests—both for the base budget and the war-related requests—submitted to the Congress this year in some historical context, as there has been, understandably, sticker shock at their combined price tags—more than $700 billion total.

Please consider that, at about 4 percent of America's Gross Domestic Product, the amount of money the United States is projected to spend on defense this year is actually a smaller percentage of GDP than when I left government 14 years ago following the end of the Cold War—and a significantly smaller percentage than during previous times of war, such as Vietnam and Korea.

Since 1993, with a defense budget that is a smaller relative share of our national wealth, the world has gotten more complicated, and arguably more dangerous. In addition to fighting the Global War on Terror, we also face:
- The danger posed by Iran's and North Korea's nuclear ambitions, and the

threat they pose not only to their neighbors, but globally, because of their

record of proliferation;
- The uncertain paths of China and Russia, which are both pursuing sophisticated military modernization programs; and
- A range of other potential flashpoints, challenges and threats.

In this strategic environment, the resources we devote to defense at this critical time should be at the level to adequately meet those challenges.

### Fiscal Year 2007 Supplemental Request

The FY 2007 Supplemental Request includes $39.3 billion to provide the incremental pay, supplies, transportation, maintenance and logistical support to conduct military operations. The additional U.S. ground and naval forces being sent to the Iraq theater are projected to cost $5.6 billion. This total includes funding for personnel costs, supplies, spare parts, contractor support, and transportation. The FY 2008 GWOT Request complies with Congress's direction to include the costs of the conflicts in Iraq and Afghanistan in the annual Defense Department budget.

### Reconstitution

The request includes $13.9 billion to reconstitute our nation's armed forces—in particular, to refit the ground forces, the Army and Marine Corps, who have borne the brunt of combat in both human and material terms. These funds will go to repair or replace equipment that has been destroyed, damaged, or stressed in the current conflict.

All Army units deployed, or about to deploy, for missions overseas are fully trained and equipped, often with additional gear for their particular mission. In an expeditionary, rotational force one can expect that units returning from their deployment will decline to a lower readiness level as personnel turn over and equipment is repaired or replaced.

### Force Protection

This supplemental includes $10.4 billion for investments in new technologies to better protect our troops from an agile and adaptive enemy. Programs being funded would include a new generation of body armor, vehicles that can better withstand explosions from Improvised Explosive Devises (IEDs), and electronic devices that interrupt the enemy's ability to attack U.S. forces. Within this force protection category, the FY 2007 Supplemental includes $2.4 billion to counter and defeat the threat posed by IEDs.

## Afghan/Iraqi Security Forces

The request includes $9.7 billion to stand up capable military and police forces in Afghanistan and Iraq.

The bulk of these funds are going to train and equip Afghan National Security Forces (ANSF) to assume the lead in operations throughout Afghanistan. Some 88,000 have been trained and equipped, an increase of 31,000 from the previous year.

The $5.9 billion for the ANSF in the FY 2007 Supplemental is a substantial increase over previous years' appropriations. It reflects the urgent priority of countering increased activity by the Taliban, Al Qaeda, and narcotics traffickers to destabilize and undermine the new democracy in Afghanistan. These funds will significantly upgrade the capability of Afghan forces to conduct independent counter-insurgency operations.

In Iraq, more than 300,000 soldiers and police have been trained and equipped, and are in charge of more than 60 percent of Iraqi territory and more than 65 percent of that country's population. They have assumed full security responsibility for three out of Iraq's 18 provinces and are scheduled to take over more territory over the course of the year. These Iraqi troops, though far from perfect, have shown that they can perform with distinction when properly led and supported. Iraqi forces will be in the lead during operations to secure Baghdad's violent neighborhoods. By significantly increasing and improving the embedding program, Iraqi forces will operate with more and better Coalition support than they had in the past.

## Non-Military Assistance

Success in the kinds of conflicts our military finds itself in today—in Iraq, or elsewhere—cannot be achieved by military means alone. The President's strategy for Iraq hinges on key programs and additional resources to improve local governance, delivery of public services, and quality of life—to get angry young men off the street and into jobs where they will be less susceptible to the appeals of insurgents or militia groups.

Commander's Emergency Response Program, or (CERP) funds are a relatively small piece of the war-related budgets—$456 million in the FY 2007 Supplemental. But because they can be dispensed quickly and applied directly to local needs, they have had a tremendous impact—far beyond the dollar value—on the ability of our troops to succeed in Iraq and Afghanistan. By building trust and confidence in Coalition forces, these CERP projects increase the flow of intelligence to commanders in the field and help turn local Iraqis and Afghans against insurgents and terrorists.

## Conclusion

With the assistance and the counsel of Congress, I believe we have the opportunity to do right by our troops and the sacrifices that they and their families have made these past few years. That means we must make the difficult choices and commit the necessary resources not only to prevail in the current conflicts in which they are engaged, but to be prepared to take on the threats that they, their children, and our nation may face in the future.

## Senior Leadership Meeting of the National Guard

*National Guard Bureau, Washington, DC, Tuesday, February 27, 2007*

THANK you, General Blum.

I appreciate the opportunity to meet with you today to describe briefly some of the ways we are trying to address the challenges and stresses faced by Guard members and their families.

I think America's "citizen soldiers" are unique in the history of armies, not just because of their patriotism, dedication and skill, but also because they are American citizens first and foremost. And, thus, they are not overly impressed with rank and they are unafraid to ask questions or offer advice or to criticize.

This dates back to the Revolutionary War. One of my favorite stories from that war is about the time General Washington was making his rounds and spotted a certain Private John Brantley drinking wine. Brantley invited Washington to tip the jug with him but the General retorted, "My boy, you have no time for drinking wine." Brantley, in turn, and probably a little drunk, responded, "Well then, damn your proud soul for being above drinking with your soldiers." Washington turned back, and said, "Come, I will drink with you." After the jug was passed and Washington re-mounted his horse to ride off, Brantley yelled after him, "Now I'll be damned if I won't spend the last drop of my heart's blood for you."

Well, I suppose our citizen soldiers are a little more disciplined than that now. But, in Iraq, Afghanistan and elsewhere, when I share a meal with our troops, they are unafraid to say what they think and to criticize. And, I hope we never change that. Because it means American democracy is planted firmly in the spirit and hearts of our citizen soldiers.

On my last visit to Iraq, I spoke with a unit from the Minnesota National Guard—a unit that recently had its tour of duty extended. Despite the difficult circumstances they were in, I was struck by their positive attitude. They were proud of what they had accomplished in a region that had once been one of Iraq's most dangerous. I was thankful to be able to tell them in person how much I appreciated everything they are doing for our country.

I would also like to acknowledge the role played by the families of Guard members—the thousands of husbands and wives, and sons and daughters, who are giving for our country as well. While their loved ones are away serving at some distant post, Guard families remain at home, dealing with day-to-day issues. They mow the lawn, shovel the snow and pay the bills; they make sure the kids get to soccer practice. And they perform the thousands of other tasks that keep their family going in the absence of their loved ones. Their quiet but noble efforts deserve our attention and gratitude.

The support of these family members—as well as their employers and communities—has been crucial to keeping our best citizen soldiers in the Guard. In fiscal year 2006, the Army Guard exceeded its retention mission by 18 percent. Even with the strain of extended deployments, and homeland security and border missions, the Army Guard was able to meet its annual goal four months early.

The willingness of patriotic young Americans to sign up for the Guard has been equally impressive, given the high likelihood of dangerous duty overseas. In the last fiscal year, the Army Guard achieved 99 percent of its recruiting goal, and actually signed up 19,000 more soldiers than in fiscal year 2005.

Due in large part to what has arguably been the most cost-effective recruiting effort in the military, the Army Guard has seen a net increase of some 14,000 soldiers over the past year.

This strong showing is a tribute to the men and women who choose to join and stay in the Guard, as well as the efforts of each of you here today. And I commend you for it.

Since coming back to Washington after nearly 14 years, I've been struck by the many changes in the way our government and military are arranged and operated. And one of the most dramatic shifts has been in the role and capabilities of the National Guard.

For much of the last century, the Guard was largely considered a strategic reserve, standing by in case of a mass mobilization. It was not a priority for funding and equipment, even though its members had served in every conflict from the Revolutionary War onwards.

Since September 11, we've seen a remarkable transformation of the Guard—from a strategic reserve to a fully operational reserve that is an integral, indeed an indispensable, part of America's pool of forces used in Iraq, Afghanistan and elsewhere in the broader Global War on Terror. The Department is committed to providing adequate resources so it remains a truly operational force.

Though the men and women of the Guard have responded to these challenges with real spirit and resilience, the high tempo of operations and frequency of deployments in recent years has created stress on the force.

Even before I came to this job, I had two concerns about the state of the U.S. military. One was that the Army and the Marine Corps were not big enough to accommodate the multiple missions that they had been given over the past dozen years or so. The second was the use and condition of the National Guard.

Since becoming Secretary, I've tried to address both of those issues in decisions that I've either made or recommended to the President.

As you know, we will be increasing the permanent end strength of the Army and Marine Corps by some 92,000 over the next five years. One of the effects of this increase, over time, should be that with a larger pool of ground forces available, it will be less necessary to call on Guard formations as often for overseas deployments.

The second major shift has been a change in the way we mobilize and use troops and units from America's Guard and Reserves.

The goal is to distribute more fairly, and more effectively, the burdens of war among our active and reserve components, while providing a more predictable schedule of mobilizations and deployments for troops, their families, and their civilian employers. Up until now, the deployment of a guardsman or reservist for one year in Iraq or Afghanistan—the standard tour length for the Army—would usually entail up to 18 months of active duty, including time for pre-deployment training and post-deployment recovery.

All reserve component personnel, including the Army National Guard, will now be mobilized for a maximum of 12 months at a time, with the goal of five years at home before their next mobilization.

We have also rescinded the policy, established in the months following the September 11th attacks, that set a cumulative limit of 24 months of being involuntarily mobilized over the course of a reservist's or guardsman's military career. I am told that one effect of that policy was that the Army was forced to cobble together Guard battalions and brigades with personnel taken from other units, and in many cases, from other states.

It is important that citizen soldiers who live together and train together will also deploy and fight together. This change is part of an overall shift in our mobilization policies away from being focused on individuals and towards improving the cohesiveness and capability of units.

The intent of these changes is to establish a cycle of one year on active duty followed by five years at home. We are not there yet. Because of the demands on our military today, some guardsmen will have to deploy sooner than they had expected or wanted. Others will serve longer than they anticipated or would like. I have ordered that we provide additional compensation for those so affected, and have directed a review of our current waiver policy for men and women who may experience undue hardships. Also, I have directed the services to minimize the use of stop-loss.

Combined with other initiatives to better organize, manage, and take care of the force, these recent changes should mean that in the future our troops should be deployed or mobilized less often, for shorter periods of time, and with more predictability and a better quality of life for themselves and their families.

As you probably know, the Commission on the National Guard and Reserves is scheduled to give its interim report on Thursday, with recommendations on the National Guard Empowerment Act introduced by Congress last year. I have been briefed on the progress of the Commission and we will analyze their recommendations as soon as we get them.

I believe the Department will agree with the Commission on many issues. Where possible, I will make changes to Department policy. For those recommendations that require legislation, I will forward a proposal to Congress this year.

Whatever the changes, it is important that we do not undermine the total force concept—where the Army and Air Guard are considered fully integrated parts of their service branches. I do not favor placing the Chief of the National Guard Bureau on the Joint Chiefs of Staff, but if his responsibilities are such that they warrant a position of four stars, I will support that change. And that is my inclination.

Before closing, I would like to say a few words on the issue of readiness, a topic I know is of some concern.

The practice of leaving equipment behind in-theater has created unique challenges for Guard units returning from overseas deployments. Unlike active duty units, these Guard formations must always be on call in the case of a domestic emergency or natural disaster. I know General Blum and his team have done some excellent work to better coordinate and share Guard assets across state lines when contingencies do arise, as happened with Hurricane Katrina.

But it is understood that mutual support agreements between states are not the long-term solution to the Guard's equipment-readiness challenges.

Reconstituting and resetting the Guard and Reserve—in particular the nation's ground forces—is a top priority for the Department of Defense and we are asking in the fiscal year 2007 and fiscal year 2008 budgets for some $9 billion to get on top of this.

The goal of this program is a National Guard that will be fully manned, fully trained and fully equipped, and fully capable of taking on a range of traditional and nontraditional missions both at home and abroad. This will ensure the Guard remains "Always Ready, Always There."

From the beginning of this conflict, we have asked a tremendous amount of our citizen soldiers. They have done everything asked of them and more. I thank them for their service, and I thank you here today for your leadership.

I would be happy to take some questions.

# NCO Breakfast

*Fort Belvoir, VA, Thursday, March 01, 2007*

THANK you for inviting me here this morning. It is nice to have this opportunity to be outside of the beltway—even if not by much.

Back in October, at a conference in Iowa, I passed along some insights from previous tours of duty in Washington D.C.—a place where so many people are lost in thought, because it's such unfamiliar territory—a place where those who travel the high road of humility encounter little traffic. Little did I suspect I would be called back here a couple of weeks later! There is an adage, "Be careful of what you ask for, you might get it." Perhaps there is a Washington corollary, "Be careful of what you don't ask for, you might get it."

It has truly been an honor to be called into service as Secretary of Defense—to serve our country, and to serve with men and women like you, during this time of great consequence. I want to thank you, personally, for what you are doing each and every day.

Before I began at CIA in the late 1960s, I served as an Air Force intelligence officer at Whiteman Air Force Base, where I briefed missile crews and the occasional General on international politics. Mercifully, this was in the age before Power Point. I served there for close to a year and a half. Of course, it only took about a day and a half to figure out who really made the military run—or at least who made us junior officers run—the Non-Commissioned Officers. So I did what my sergeant suggested and the two of us did my job pretty well. Now, as then, I greatly appreciate all that you are doing.

Your job is difficult. You spend many long hours at work. There is a natural pull to be out shoulder-to-shoulder with those standing watch on ships or at forward operating bases as many of you have been these past few years. Though you may be working thousands of miles away, your impact—and your positive leadership—is being felt in every fighting hole and on every front line. You also help ensure service members have the supplies and equipment they need and the quality of life they deserve. I have had the opportunity to visit with service members on two of those front lines—Afghanistan and

Iraq. They are working in difficult conditions. When I visited Forward Operating Base Tillman along the border between Afghanistan and Pakistan, it reminded many of us of a frontier town in the Old West. The area surrounding the base was desolate and forbidding. But morale among troops in Iraq and Afghanistan is high. They are focused on their mission.

And they are proud to be serving during this moment in history. I think everyone comes away inspired by the sheer determination and professionalism of those serving today. As a mentor and a motivator, you are largely responsible for setting those marks. You should be proud of this legacy.

As you probably know, we are in the heart of the budget testimony season. This year's budget request continues to focus on improving the standard of living of service members.

For example, it includes nearly $3 billion to continue improving enlisted barracks and housing. We will privatize nearly 3,000 more new family units. I understand this base was the first to build a town center-like community as part of on-base housing. This year, the Department is building more child development centers and schools, so that we continue to take care of our families as well as our troops.

On another issue close to home, I am concerned by the reports of substandard conditions at Walter Reed hospital. Being in the military is like being part of a family. It appears that some of our family may not have been treated the way they should have been.

I suspect that each one of you here has taken the opportunity to meet with wounded troops at Walter Reed or other hospitals—likely many times. Those troops have given their all for our nation. They deserve the very best our nation can give them. The special, bipartisan review group I announced last Friday will ensure that Walter Reed and similar facilities are kept to that standard.

Let me be clear: Any individual, regardless of rank—be they officer or enlisted, military or civilian—will be held accountable for allowing this unacceptable situation to develop.

I want to mention one more thing before taking your questions.

For anyone working in this town, there is no escaping the debate taking place over the direction of this war. This should not surprise us, however. Such debates have taken place in our country since before we were even a country. If you can believe it, there were many northerners vehemently opposed to fighting the Civil War. President Lincoln nearly lost reelection because of them. I remember in 1982 when President Reagan pledged that the United States would support people fighting communism wherever they were. His words were labeled the "Reagan Doctrine" by some, and "ridiculous" by others.

But there are two important things to keep in mind. First, during our nation's most difficult times, there have been folks like you willing to stand for what is right—no matter the challenges or consequences—until their missions were complete. They helped make our nation a force for good in the world. Second, President Bush, General Pace and I—along with millions of Americans—are determined to secure the equipment, supplies and support you need so that you can complete your missions—so that our nation will remain a force for good in the generations to come.

Thank you, and I would be happy to take some questions.

## U.S. Senate Youth Program

*Pentagon Auditorium, Washington, DC, Friday, March 09, 2007*

GOOD morning. On behalf of the nearly 23,000 employees who work here at the Pentagon, and the three million civilians and military personnel in the Department of Defense, I'd like to welcome you. Kara, thank you for your kind words. To you and David, who we'll hear from later, please convey my special thanks and appreciation to your parents for their service and sacrifice.

I congratulate all of you on your selection as delegates to the United States Senate Youth Program, now in its 45th year. You have earned this highly competitive and prestigious honor. Your state school boards back home believed in you, and the U.S. Senate has chimed in with its support. I hope the visit to Washington has met your expectations so far.

The military is a strong supporter of your program—as evidenced by the 16 military officers who are here today, your mentors and ambassadors of the armed forces. I want to mention someone who was selected to be a mentor and part of this program, Captain Jennifer Harris of the United States Marine Corps. Captain Harris was a helicopter pilot, killed in the line of duty in Iraq on February 7. We honor her service, and express our condolences to her family.

All of you in the Senate Youth Program have performed at the highest level academically, and are just as accomplished in your extracurricular activities. Some of us took a little longer to learn how to multi-task, I must say. During my first semester at William and Mary, I made a D in calculus—the only D I ever got. My father called me—those were the days when your parents actually got your grades in college—my father called and said, "Tell me about the D." To which I responded, "Dad, the D was a gift." I'm guessing not many of you have had a conversation like that.

I used to tell the freshmen at Texas A&M that I had two messages for them. One, the fact that I got a D means that you can be reasonably smart and if you don't work hard you'll still get a bad grade. And the other is that you can get a bad grade and still go on to be somewhat successful.

Well, as you take your tour of official Washington and learn about U.S. government institutions, it might be good to pause for a moment and ponder this question: "How will my progress shape the future of my community, state, and country?"

You are about to become adults and take that first step toward fulfilling your dreams: the choice of whether to go to college, and if so, which one. Now of course I have to put in a plug for Texas A&M, the sixth largest university in the nation, 46,000 students. Four months ago I was president of a university. I had really bad timing. I left Texas A&M just as, for the first time in four years, the football team was nationally ranked, and just as our football team beat the University of Texas in their home stadium for the first time in 12 years, and just as our men's and women's basketball teams were nationally ranked for the first time ever. Of course I used that opportunity simply to say, "My work here is done, it's time for me to leave." But A&M has a special spirit and traditions that are very—that are unique—and I encourage you to give it a look. In fact, one of the real downsides of becoming Secretary of Defense was leaving just as we were starting half a billion dollars in academic construction and the athletic program was taking off.

Nevertheless, a university degree, wherever you earn it, has never been more valued or more valuable. The education you are pursuing will certainly have a financial payoff later in your life. But the true value of education is its ability to help you harness your knowledge and your passion to help make a difference in the lives of others.

After college, I served in the Air Force, and then joined CIA as an analyst, the first of a number of jobs in the national security arena. There are many different roles that you can fulfill, in and out of uniform. But I want to tell you, we need smart people who will grapple with the challenges of the 21st century here in the Defense Department. We also think you can serve in the State Department, in the Agency for International Development, in CIA, and in a number of other places.

But I would tell you that there is a special need and a duty at this time in the history of our nation for the best and brightest to put their talents to work in our armed forces, and I encourage you to consider military service.

America is truly blessed by the dedicated men and women who have stepped forward to raise their right hand in our nation's defense.

These young people make up as diverse, well-educated, and talented a group as this country has ever seen. The sifting and selecting that take place in all of the services are the mechanisms of a system in which merit and integrity are the strongest criteria. Every service member—each judged by what he or she is and does—wears the uniform with the pride of representing the United States.

When I was president of A&M, I interacted a fair amount with students preparing to go into the armed forces. A&M still commissions more officers into the armed forces than anyplace else in the country other than the three service academies. And during our early morning runs together, or in conversations between classes, I always felt that they carried an extra sense of pride. As Secretary of Defense, I have traveled to Iraq and Afghanistan and met with many of our troops—skilled and dedicated individuals who each day are facing dangers and stresses of combat and separation from their families.

Among the service members I've met have been National Guard members home from Iraq who volunteered to remain on active duty to help rebuild in the wake of Hurricane Katrina. Others have volunteered for additional tours in combat areas so that they could, in their own words, "make a difference."

One reads and hears often about "supporting the troops," but often people who want to help don't know where to start. A few years ago, the Department of Defense launched a program called "America Supports You," that recognizes all the things the American people do across the country to support the men and women of the U.S. military and their families. If you go to the "America Supports You" website, you can find countless projects to join—from sending care packages to soldiers deployed abroad, to lending a helping hand to their families here at home.

In closing, I want to reinforce something you already know: that character and integrity must be at the core of your quest for high achievement. These are what make a decent and productive life possible, whether in public service or in the private sector. At a time when we are all too aware of the social, political, and even economic costs of politicians who lie, business leaders who cheat and steal, and other fraudulent individuals, your honesty and your integrity will be your greatest assets. They will shine like a beacon in a storm. Develop them. Refine them. Bind yourself to them.

There was an actor in Westerns that I'm sure you're familiar with—maybe, maybe not. It may be an age thing. But one of my favorites was always John Wayne. And John Wayne was something of a personal hero of mine. And he once said in a movie that I think still holds true in real life: "There's right and there's wrong. You get to do one or the other. You do one, and you're living. You do the other and you may be walking around, but you're as dead as a beaver hat."

In holding steadfast to your honesty and your integrity, you set yourself apart. You make it inevitable that, whatever you decide to do in life, you have ahead of you a future of leadership, service, and impact.

I wish you all the best in your future endeavors, hope you have a great experience to take home with you from the nation's capital.

# U.S. Northern Command Change of Command Ceremony

*Peterson Air Force Base Colorado Springs, Colorado, Friday, March 23, 2007*

THANK you for that introduction.

Senator Allard, thank you for being here. It's good to see the members of the diplomatic corps and civic and community leaders who are with us today.

A special welcome to the members of the Keating and Renuart families.

It's an honor to be here to welcome General Renuart to NORTHCOM and NORAD—and an honor to recognize Admiral Keating's achievements in this command before he takes his next assignment.

NORTHCOM has come a long way in only a few years.

When I left government 14 years ago this command center didn't even exist—and few people were thinking seriously about the types of threats we face today.

As an old cold warrior, however, I was well aware of NORAD's vital role in our strategic nuclear defense.

In fact, not many people know that my first job out of college was at a ballistic-missile facility—Whiteman Air Force base in Missouri. Because of my academic background and moderate Russian language skills, I was assigned to do briefings on our wing's strategic minuteman missile targets, which brought me into early contact with high-ranking officers.

I recall one briefing in particular. I was explaining our target set to an Air Force lieutenant general—who I would characterize as a cigar-chomping Curtis LeMay wannabe.

When I told him that 120 of our 150 missiles were aimed at Soviet ICBMs, he immediately blew up and, with many expletives I will delete, said it was an outrage that we would be hitting only empty silos—instead of killing Russians. He demanded that I, Second Lieutenant Robert Gates, rewrite the nuclear targeting plan. Never has so much been asked of so few with so little ability to carry out the order.

Times have certainly changed—as have our military's flag officers.

In a speech last year, Admiral Keating talked about the awesome responsibility that our young men and women in uniform have in today's strategic environment. They are, he said, "warrior diplomats," the embodiment not only of America's foreign policy, but of our nation's values. I would certainly second that.

But I would add that the values of the American military—honor, duty, patriotism—start at the top. They start with men like Tim Keating, who likes to say, "In defending our homeland we find the better we do our job, the less likely folks will notice—and that's how it should be." That speaks to an ethos of grace and humility that has guided Admiral Keating throughout his career.

Admiral Keating has had numerous key assignments—commanding a carrier group in the Pacific, leading U.S. Navy forces in the Middle East—but I think he would tell you that leading NORTHCOM has been his most important.

We learned on September 11th that the fundamental nature of man has not changed, that we still live in a dangerous world, and so we must remain vigilant and prepared.

The U.S. Northern Command, which was stood up in the wake of the 9/11 attacks, acts as a command center responsible for protecting the security and integrity of the homeland. NORTHCOM's duties include:

- Planning for and responding to natural disasters;
- Missile defense;
- Border security; and
- Coordinating to protect against terrorist attacks at events ranging from the State of the Union address to the Superbowl.

It is a difficult balancing act to protect the U.S. from external threats as well as deal with internal emergencies—and to do so within the bounds of our laws and Constitution.

Under Admiral Keating's leadership, NORTHCOM has done both well.

Consider this command's response to Hurricane Katrina, which devastated some 90,000 square miles, an area roughly the size of Great Britain. Within hours of the hurricane's landfall, NORTHCOM was already running operations—operations that quickly grew to 22,000 active-duty servicemen in theater; more than 300 helicopters doing search and rescue; and at least 20 ships.

The ability to direct complex missions, and especially the ability to coordinate among many competing interests, will be of special importance at Pacific Command, an area that includes more than 70 nations and territories. Of course, Tim knows a thing or two about diplomacy. Having to deal with 50 governors and their Adjutants General is at times like dealing with 50 miniature heads of state, commanders-in-chief, and defense secretaries all rolled into one.

I would like to mention a group of people that doesn't receive as much publicity as they should—the families of men like Admiral Keating. This is a difficult job in the best of times, and it would be impossible without the understanding and support of a loving family.

Wandalee, thank you for being there all these years. I know everyone at NORTHCOM will miss you.

I also want to thank Admiral Keating's children, Dan and Julie, and of course the grandchildren, Lauren Joy and Matthew.

Dan, and Julie's husband Paul, are both F-18 pilots. And both have flown combat missions in support of Operation Iraqi Freedom. They certainly have a good role model in Admiral Keating—but they have a lot of work ahead of them if they want to match his 5,000 flight hours and 1,200 arrested landings.

We are truly blessed to have a family that has given so much to our nation. Admiral Keating, thank you for your service.

I wish you and your family "fair winds and following seas."

And, as luck would have it, I get to extol Admiral Keating's virtues all over again a week from now in Hawaii to mark his assumption of Pacific Command.

Admiral Keating's successor, General Renuart, is well prepared to lead this command. He has seen combat on the ground and from afar, and has proven to be one of the military's most seasoned and effective leaders.

He commanded the 76th Fighter squadron during the first Gulf War, and has flown combat missions in operations Desert Storm, Deny Flight, Northern Watch, and Southern Watch.

One of his staffers mentioned that the reason Gene is so proud of his Audi is probably because it's faster than his plane of choice—the A-10 Warthog, regarded as the slowest, ugliest plane in the Air Force, and the best friend a soldier or Marine could have in a close fight.

More recently, Gene served as director of operations for Central Command during the planning and execution of the conflicts in Afghanistan and Iraq, a position where he oversaw all joint and allied combat missions, humanitarian assistance, and reconstruction operations.

He has also served as director of strategic plans and policy for the Joint Staff. In that capacity, he worked closely with the NSC and the State Department to offer direction and guidance on a variety of issues, including:

- Building the capacities of partner nations;
- Strategic communications and outreach;
- International negotiations; and

- Many of the "long war" issues that are so important in the ongoing battle against jihadist terror both in the Middle East and across the globe.

Aside from that, with all the personal memorabilia he's managed to fit into his office next to mine at the Pentagon, I think he officially curates the smallest United States Air Force museum in the nation.

Gene, I hope all this talk about the challenges and responsibilities of NORTHCOM hasn't scared you off. Of course, if your most recent assignment as senior military assistant to the Secretary of Defense hasn't done that, I doubt anything will.

Thank you for your help in getting me acclimated to life at the Pentagon. I wish you the best of luck with your new assignment. The security of our homeland is in good hands.

Thank you.

# U.S. Pacific Command Assumption of Command Ceremony

*Camp H.M. Smith, Hawaii, Monday, March 26, 2007*

GOOD morning. As President of Texas A&M University, I used to begin all of my speeches by saying "howdy." It is a pleasure to be here this morning, and to be able to say "aloha."

I want to first thank the men and women of U.S. Pacific Command. I know that Admiral Keating would agree that much of what has been accomplished in this region is due to your dedication and professionalism. You and your families have our appreciation and thanks for everything you do for our nation.

After spending last night and this morning around this area, it is easy see why many of you chose to come here. Not only are you surrounded by such natural beauty, but you also get to be five time zones away from Washington, D.C.

I hope that Governor Lingle and General Pace will vouch for me if I take a little extra time returning to Washington.

Welcome General Chilton. And welcome to three former leaders of Pacific Command: Admiral Hayes, Admiral Fargo, and Admiral Macke. Thank you all for being here.

I appreciate the representation today from countries throughout the Pacific Command region—a testament to the partnerships that have been fostered under Admiral Fallon. Welcome, and thank you for being here. I suspect that Admiral Keating may be visiting many of your countries in the coming months.

Wandalee, it is good to see you again. Three days ago, I had the opportunity to honor Admiral Tim Keating and Wandalee as they left U.S. Northern Command. I am pleased to have that opportunity again today. You two will be sure to stop me if you have heard all this before.

As I noted when recommending Admiral Keating for this post, he has established a record of accomplishment in a variety of complex and challenging assignments. He commanded a carrier group based in Japan, and later the Navy's Fifth Fleet during Operation Iraqi Freedom. In his most recent post as the head of Northern Command and NORAD, he was responsible for guarding our homeland against a range of threats and means of attack from weapons of all kinds—some so small that they could even fit inside a thimble. His area of responsibility ranged from our nation's cities and coastlines to outer space and everything in between. These responsibilities entailed working with local, state, and federal officials and politicians including 50 governors and adjutants general.

So perhaps—just perhaps—Admiral Keating is one of only a few who would find taking on Pacific Command's region—encompassing more than 50 percent of the earth's surface and 43 countries—slightly less daunting.

He may also find it less daunting because he served here before as a Flag Lieutenant to the head of Pacific Command. He may be one of only a few such aides to ever assume the post he once served.

Admiral Keating's career has also been groundbreaking in that this will be his sixth joint tour. Though I understand that during the week before the Army-Navy football game—or as Tim would probably call it, the Navy-Army game—he is anything but joint. He wears a ball cap proudly displaying "Beat Army." Last year, some unknown individual wrote "Go Army, Beat Navy!" on a mirror near his office.

Tim, I will be keeping a close eye on movements of the Navy's Seventh Fleet in case there are any similar incidents this year.

That one week a year aside, Admiral Keating has proven himself to have the right mix of military, intellectual—and yes—diplomatic skills to excel at this vital post.

Pacific Command's sheer size demands any country's attention. But more than that, the scope of relationships and challenges encompassed in this area of responsibility are key to the security and continued welfare of the United States. It is home to some of America's oldest and strongest allies—and to some relatively new relationships as well. A great many partnerships across this Command—old and new—have grown considerably stronger in recent years. The restoration of military relations with Indonesia comes to mind, as does the strengthening of our long-standing ties with Japan and Australia.

Part of this is due to Admiral Fox Fallon's legacy. He built a level of trust and cooperation among nations that I am confident Admiral Keating will continue to foster. But it is also due to a growing realization that we share common challenges.

Together, nations across the region face the newest of threats in the proliferation of weapons of mass destruction technology. We face the oldest of threats from piracy combined with dangerous new technologies and materials in the Straits of Malacca and elsewhere. Countries with limited transparency are taking actions that seem contrary to international stability—causing other countries to question their intentions. And violent jihadists are trying to undermine the foundations of free society that have allowed many countries in this region to prosper.

Admiral Keating fully understands the challenges of today. During numerous tours and deployments across the Pacific and Indian Oceans, he has developed a keen appreciation for what he calls "the vibrancy and complexity of this vast region."

He is fully committed to defeating the enemies who threaten our security and stability, because he has seen first-hand the consequences of failure. He was in the Pentagon when it was struck by extremists on September 11th. Some of those who died that morning worked for him.

And he is fully prepared to continue the record of accomplishments that the countries of this region have built together. In nearly two and a half years at Northern Command, he worked tirelessly to create many new relationships after the formation of Northcom, and he strengthened existing partnerships to increase our nation's security.

Tim, congratulations on this assignment. We wish you and Wandalee all the best.

# American-Turkish Council

*Washington, DC, Tuesday, March 27, 2007*

THANK you, Tom, for that kind introduction.

And Brent, thank you for those uncharacteristically generous remarks.

Brent Scowcroft is someone I have known and worked with since I was first detailed for the National Security Council from CIA 33 years ago this summer. Brent was then the deputy national security adviser. And as I recall, President Nixon's final appeal for the Watergate case was being heard by the Supreme Court. I later wrote that working for Brent in the White House at that time was like being a deckhand on the Titanic.

When Brent offered me the deputy national security adviser job 15 years later, in 1989, I accepted on one condition—that I would keep his work schedule, which averaged 14 to 16 hours a day. I had two young children, and I wanted to see them once in a while. Brent quickly agreed to my conditions, believing clearly that I would never stick to them. I didn't. He knew me better than I knew myself.

Even that occasion was far from the last time he talked me into doing something I didn't want to do, which would then turn out to be a life-changing experience for me. But even Brent didn't have the temerity to try to talk me into taking my present job.

He did offer some guidance for me today. It was along the lines of, "Remember, there's no such thing as a bad short speech."

That was kinder than the time George Bernard Shaw was introducing a speaker and told him he only had 15 minutes to speak. The speaker replied, "How can I tell them what I know in 15 minutes?" And Shaw responded, "I advise you to speak very slowly."

It's been just over three months since I returned to Washington to take on my current assignment. You know, the place—the Pentagon is just a huge place. David Brinkley told a story about a woman who told a Pentagon guard she was in labor and needed help in getting to a hospital. The guard said, "Madame, you shouldn't have come in here in that condition." She replied, "When I came in here, I wasn't."

But seriously, in a relatively short period of time I've had the opportunity to meet and work with some extraordinary people dedicated to serving their country.

At the top of that list are our men and women in uniform, and especially those serving on the front lines in Iraq and Afghanistan. I've come away from my visits in theater impressed by their resilience, their good humor, their courage and their determination in the face of danger and personal sacrifice.

At the Pentagon, one of the more impressive people with whom I've become reacquainted is no stranger to this group—former U.S. Ambassador to Turkey, and current Under Secretary for Policy Eric Edelman. As is often the case, Eric is on the road today doing the nation's business. I'm sure he won't mind if I tell you that, as a career diplomat, he manages to do very well in a building where more than a few share Will Rogers' view that diplomacy is the art of saying "nice doggy" until you find a rock.

One of the first conversations I had with Eric on my return to Washington was about the importance of America's relationship with Turkey, an ally my friends know I have long believed to be undervalued and under-appreciated. And for that reason, as the Soviets used to say, it is not by accident that my first public speech since becoming secretary is before this group.

Today, I'd like to speak a few minutes about that relationship and about the challenges to our shared values and interests in the Middle East in the years ahead.

It was 60 years ago this month that President Harry Truman addressed a joint session of the U.S. Congress to ask for emergency assistance for Greece and Turkey to stave off Soviet expansion in the Mediterranean area.

In those historic remarks—subsequently known as the "Truman Doctrine" speech, President Truman laid out the fundamental objectives of U.S. foreign policy for the next four decades, to include, quote, "the creation of conditions in which we and other nations will be able to work out a way of life free from coercion," unquote. A central component of that policy was maintaining Turkey's "national integrity," which he said was "essential to the preservation of order in the Middle East."

That holds true today. In addition to the commitments we have made to each other as long-standing allies, there is Turkey's unique cultural and geographic position in the world—a position of vital importance to the security challenges we face today.

Rudyard Kipling wrote that, "East is East, and West is West, and never the twain shall meet," but in fact, Kipling was wrong because for centuries Turkey has served as a bridge between East and West.

Many of the competing forces that historically have been at play in Turkey are also found in the Middle East in the broader Muslim world—extremism versus moderation; tribalism versus nationalism; authoritarianism versus pluralism; and secularism versus Islamism.

Turkey has been the seed of several civilizations—including the Byzantine and Ottoman empires. In the last half of the 20th century, Turkey courageously defended NATO's southern border against a very different empire, helping hold off Soviet aggression during some of the most difficult days of the Cold War.

For nearly 200 years, Turkey has been on a modernizing and westward course—a course set early by Ottoman reformers in the 19th century, strengthened and given modern form by Kemal Ataturk, and continued by men like Turgut Ozal. It is a course the United States has strongly favored and continues to support.

It is no secret that the strategic relationship between the United States and Turkey has undergone some turbulence in recent years. Even so, our military, economic, political, and personal ties remain strong.

Turkey is, for example, one of the major allied partners on the Joint Strike Fighter, and 16 U.S. Navy ships called on Turkish ports last year.

Turkey provides key support in the difficult struggles against violent extremism we find ourselves in today:

- America appreciates the role Turkey has played in the effort to help the Afghan people build a secure and stable democracy. Turkey has commanded two ISAF rotations and a Provincial Reconstruction Team.
- Turkey has provided access to Iraq through the Incirlik Air Base and the Habur gate, without which our operations would be exceedingly more difficult and vastly more expensive.

Regardless of the resilience of our strategic relationship, we should not forget that all relationships need work to remain strong. Our two nations should oppose measures and rhetoric that needlessly and destructively antagonize each other.

That includes symbolic resolutions by the United States Congress, as well as the type of anti-American and extremist rhetoric that sometimes finds a home in Turkey's political discourse.

No doubt one of the most difficult matters we have had to work through as allies is the conflict in Iraq. Indeed, it is perhaps the most difficult matter facing many allies of the United States.

The situation on Turkey's border with Iraq's Kurdish region is of particular concern. We recognize that every Turkish citizen killed by the PKK is a setback for success in Iraq and a setback in our relationship with Turkey. The United States has appointed one of our most distinguished military officers, General Joe Ralston—former NATO commander—as Special Envoy for Countering the PKK. But we know more needs to be done.

As President Bush has underscored, the United States is committed to the stability and territorial integrity of Iraq, and opposes policies or groups that would undermine that integrity in any way.

As I said to some of our allies and friends last month in Munich, whatever disagreements we might have over how we got to this point in Iraq, the consequences of a failed state in Iraq—of chaos there—will adversely affect every member of the Atlantic Alliance—and none more so than Turkey.

In Iraq, we face historical rivalries and rifts in a society brutalized by decades of dictatorship and war—and more recently by terrorist attacks and sectarian violence. The Iraqis are trying to do something that has never been done before in their long history—create a government that actually serves the people. And the elected government they now have essentially started from scratch less than a year ago.

The Coalition has a new commander—General David Petraeus—who's implementing a new approach based on sound counterinsurgency principles. Providing basic security and a decent quality of life for the population are the top priorities—an approach that gives the Iraqi government breathing room to take steps toward reform and political reconciliation.

Additional U.S. and Iraqi forces are being employed to "hold" and "build" neighborhoods that have been cleared of insurgents, militia, and foreign terrorists. The Iraqi government has committed to providing the forces necessary to secure their capital, and there are encouraging signs, though it is too early to call them trends.

But as General Petraeus has said, the situation will not, and cannot, be resolved ultimately by military means.

Iraq's neighbors will need to play a constructive role going forward, even if they haven't done so in the past—especially in encouraging political reconciliation and a reduction in violence within Iraq.

This is certainly the case with Syria and Iran, who have not been helpful. The regional talks recently held in Baghdad were a good start toward improved cooperation, and our government is open to higher-level exchanges.

It is no secret that, as a private citizen, I advocated for some dialogue with Iran. But in dealing with a regime like Iran's, one has to be realistic.

The American search for elusive Iranian "moderates" is a recurring—and mostly fruitless—theme since the revolution in 1979. I remember when then-National Security Adviser Zbigniew Brezinsky and I met in Algiers along the with the Iranian political leadership on November 1st, 1979. Brezinsky offered American cooperation and recognition of the Iranian Revolutionary Government.

He offered to continue the partnership that had previously existed under the shah, including military assistance to the new revolutionary government. Our interlocutors had only demand: give us the Shah. And ultimately, Zbig said that would be incompatible with our national honor. Three days later, word came that 66 of our diplomats had been seized at our embassy in Tehran, and two weeks later, the president—the prime minister, the Defense minister and the foreign minister, with whom we had met, were all out of office, and several of them in jail.

We should have no illusions about the nature of this regime or about their designs for their nuclear program, their intentions for Iraq or their ambitions in the Gulf region.

Still, at this time, Iran and all the actors in the region, friends and adversaries alike, are invested and involved to some degree or another in what is happening in Iraq.

They are watching what the United States and our coalition partners are doing and will draw their own conclusions about the reliability of our word and the strength of our commitments.

Multiple administrations of both American political parties have concluded that stability in the Gulf region is a vital American interest—an interest and a responsibility we will not abandon.

America's approach to this part of the world has over time lent itself to discussions over conflicting U.S. policy traditions and schools of thought. There are debates over the role of:
- Realism versus Idealism,
- Stability versus Freedom; and
- Interests versus Values.

In the real world, I believe American foreign policy must be a blend of all these approaches, with different emphases in different places and at different times. What matters are results that benefit the long- term security, prosperity, freedom, reputation and influence of the United States and our allies.

Abandoning Iraq and leaving regional chaos in our wake clearly would be an offense to our interests as well as our values, a setback for the cause of freedom as well as the goal of stability.

The realities we face today in the Middle East and elsewhere are a stark reminder of the fact that the fundamental nature of man has not changed. There will always be those who will not bow to reason, to accommodation or restraint—those who have few, if any, "better angels" to whom we might appeal.

I remember what Harold Brown, Secretary of Defense during the Carter Administration, used to say about the arms race with the former Soviet Union. He said: "when we build, they build; when we cut, they build."

This dynamic was true during the Cold War, and it is the case today in dealing with the jihadist terrorism and other bad actors.

In this strategic environment, we have to be willing to spend the resources, absorb the costs, take the risks and meet the commitments we make to one another. It means having the credibility, ingenuity and skill to dissuade and divide our potential adversaries, while reassuring and uniting our friends.

These principles apply to our new partners—like the brave Iraqis and Afghans who are risking their lives to build, in President Truman's words, "a way of life free from coercion." And these principles apply as well to decades-old friendships like the Turkish- American relationship that brings us here together today.

And so I would like to close with an old Turkish proverb that we should keep in mind as our two nations face together a variety of threats and challenges as well as agreements and disagreements in the years ahead, and that is, quote, "A wise man remembers his friends at all times, a fool, only when he has need of them." The United States and Turkey have wisely remembered our friendship at all times.

Thank you very much.

## House Appropriations Committee—Defense

*Washington, DC, Thursday, March 29, 2007*

CHAIRMAN Murtha, Congressman Young, members of the Committee:

I thank the Committee for all you have done to support our military these many years, and I appreciate the opportunity to provide an overview of the way ahead at the Department of Defense through the President's Fiscal Year 2008 Defense Budget, which includes the base budget request and the FY 2008 Global War on Terror Request.

As to the President's defense budget requests, I believe it is important to consider their combined price tag—more than $700 billion—in some historical context as there has been, understandably, some element of sticker shock at the total.

But consider that, at about 4 percent of America's Gross Domestic Product, the amount of money the United States is expected to spend on defense this year is actually a smaller percentage of GDP than when I left government 14 years ago, following the end of the Cold War—and a significantly smaller percentage than during previous times of war, such as Vietnam and Korea.

Since 1993, with a defense budget that is a smaller relative share of our national wealth, the world has gotten more complicated, and arguably more dangerous. In addition to fighting the Global War on Terror, we also face:

• The danger posed by Iran's and North Korea's nuclear ambitions, and the threat they pose not only to their neighbors, but globally, because of their records of proliferation;
• The uncertain paths of China and Russia, which are both pursuing sophisticated military modernization programs; and
• A range of other potential flashpoints and challenges.

In this strategic environment, the resources we devote to defense should be at the level to adequately meet those challenges.

Five times over the past 90 years the United States has either slashed defense spending or disarmed outright in the mistaken belief that the nature of man or behavior of nations had somehow changed, or that we would no longer need capable, well funded military forces on hand to confront threats to our nation's interests and security. Each time we have paid a price.

The costs of defending our nation are high. The only thing costlier, ultimately, would be to fail to commit the resources necessary to defend our interests around the world, and to fail to prepare for the inevitable threats of the future.

As Sun Tzu said more than 2,500 years ago, "The art of war teaches us to rely not on the likelihood of the enemy's not coming, but on our own readiness to receive him; not on the chance of his not attacking, but rather on the fact that we have made our position unassailable."

This holds true for our national defense today.

## FY 2008 Base Budget

The President's FY 2008 base budget request of $481.4 billion is an increase of 11.4 percent over the projected enacted level of FY 2007, and provides the resources needed to man, organize, train, and equip the Armed Forces of the United States. This budget continues efforts to reform and transform our military establishment to be more agile, adaptive, and expeditionary to deal with a range of both conventional and irregular threats.

Some military leaders have argued that while our forces can support current operations in the War on Terror, these operations are increasing risks associated with being called on to undertake a major conventional conflict elsewhere around the world. This budget provides additional resources to mitigate those risks.

The FY 2008 base budget includes increases of about $16.8 billion over last year for investments in additional training, equipment repair and replacement, and intelligence and support. It provides increases in combat training rotations, sustains air crew training, and increases ship steaming days.

## Increase Ground Forces

Despite significant improvements in the way our military is organized and operated, the ongoing conflicts in Iraq and Afghanistan have put stress on parts of our nation's ground forces.

In January, the President called for an increase in the permanent active end strength of the Army and Marine Corps of some 92,000 troops by FY 2012. The base budget request adds $12.1 billion to increase ground forces in the next fiscal year, which will consist of 7,000 additional Soldiers and 5,000 additional Marines.

Special Operations Forces, who have come to play an essential and unique role in operations against terrorist networks, will also grow by 5,575 troops between FY 2007 and FY 2008.

## Strategic Investments—Modernization

The base budget invests $177 billion in procurement and research and development that includes major investments in the next generation of technologies. The major weapons systems include:
- Future Combat System ($3.7 billion)—The first comprehensive modernization program for the Army in a generation.
- Joint Strike Fighter ($6.1 billion)—This next generation strike aircraft has variants for the Air Force, the Navy, and the Marine Corps. Eight international partners are contributing to the JSF's development and production.
- F-22A ($4.6 billion)—Twenty additional aircraft will be procured in FY 2008.
- Shipbuilding ($14.4 billion)—The increase of $3.2 billion over last year is primarily for the next generation aircraft carrier, the CVN-21, and the LPD-17 amphibious transport ship. The long-term goal is a 313-ship Navy by 2020.

## Missile Defense

I have believed since the Reagan administration that if we can develop a missile defense capability, it would be a mistake for us not to do so. There are many countries that either have or are developing ballistic missiles, and there are at least two or three others—including North Korea—that are developing longer-range systems. We also have an obligation to our allies, some of whom have signed on as partners in this effort. The department is proceeding with negotiations with Poland and the Czech Republic on establishing a missile defense capability in Europe while we work with our other allies, including the United Kingdom, on upgrading early warning radar systems. We are willing to work with others in developing this defensive capability, including Russia. The missile defense program funded by this request will continue to test our capability against more complex and realistic scenarios. I urge the committee to approve the full $9.9 billion requested for the missile defense and Patriot missile programs.

## Space Capabilities

The recent test of an anti-satellite weapon by China underscored the need to continue to develop capabilities in space. The policy of the U.S. Government in this area remains consistent with the long-standing principles that were established during the Eisenhower administration, such as the right of free passage and the use of space for peaceful purposes. Space programs are essential to the U.S. military's communications, surveillance, and reconnaissance capabilities. The base budget requests about $6.0 billion to continue the development and fielding of systems that will maintain U.S. supremacy while ensuring unfettered, reliable, and secure access to space.

## Recapitalization

A major challenge facing our military is that several key capabilities are aging and long overdue for being replaced. The prime example is the Air Force KC-135 tanker fleet, with planes that average 45 years of age, the fleet is becoming more expensive to maintain and less reliable to operate. The Air Force has resumed a transparent and competitive replacement program to recapitalize this fleet with the KC-X aircraft. The KC-X will be able to carry cargo and passengers and will be equipped with defensive systems. It is the U.S. Transportation Command's and the Air Force's top acquisition and recapitalization priority.

## Quality of Life—Sustaining the All-Volunteer Force

Our nation is fortunate that so many talented and patriotic young people have stepped forward to serve, and that so many of them have chosen to continue to serve. So far, all active branches of the U.S. military exceeded their recruiting goals, with particularly strong showings by the Army and Marine Corps. The FY 2008 request includes $4.3 billion for recruiting and retention to ensure that the military continues to attract and retain the people we need to grow the ground forces and defend the interests of the United States.

We will continue to support the all-volunteer force and their families through a variety of programs and initiatives. The budget includes:
- $38.7 billion for health care for both active and retired service members;
- $15 billion for Basic Allowance for Housing to ensure that, on average, troops are not forced to incur out-of-pocket costs to pay for housing;
- $2.9 billion to improve barracks and family housing and privatize an additional 2,870 new family units; and
- $2.1 billion for a 3 percent pay increase for military members.

In addition, recently announced changes in the way the military uses and employs the Reserves and National Guard should allow for a less frequent and more predictable mobilization schedule for our citizen soldiers.

Combined with other initiatives to better organize, manage, and take care of the force, these changes should mean that in the future our troops should be deployed or mobilized less often, for shorter periods of time, and with more predictability and a better quality of life for themselves and their families.

## Train and Equip Authorities

Building the capacity and capability of partners and allies to better secure and govern their own countries is a central task in the larger war on terrorism. It is much better for partner countries, rather than U.S. forces, to deal with the terrorist networks within their borders.

In recent years we have struggled to overcome the patchwork of authorities and regulations that were put in place during a very different era—the Cold War—to confront a notably different set of threats and challenges.

The Congress took a farsighted step to overcome these impediments with the creation of Section 1206 authority, which allows the Defense and State Departments to rapidly and effectively train and equip partner military forces.

We are seeking dedicated funding of $500 million in the FY 2008 base budget and $300 million in the Supplemental to provide the combatant commanders with the resources to implement this authority. This is a joint enterprise with the Department of State that is becoming a model capacity-building effort for the Long War. The Chairman of the Joint Chiefs and the combatant commanders regard this program as the most important authority the military has to fight the war on terror beyond Iraq and Afghanistan. It allows us to help others get ahead of threats, exploit opportunities, and reduce stress on our servicemen and women.

## Global War on Terror Request

The President's FY 2008 Global War on Terror Request for $141.7 billion complies with Congress's direction to include the costs of ongoing operations in Iraq and Afghanistan in the annual Defense Department budget. Given the uncertainty of projecting the cost of operations so far in the future, the funds sought for the FY 2008 GWOT Request are generally based on a straight-line projection of current costs for Iraq and Afghanistan. This request includes $70.6 billion to provide the incremental pay, supplies, transportation, maintenance and logistical support to conduct military operations.

## Reconstitution

The FY 2008 GWOT Request includes $37.6 billion to reconstitute our nation's armed forces—in particular, to refit the ground forces, the Army and Marine Corps, who have borne the brunt of combat in both human and material terms. These funds will go to repair or replace equipment that has been destroyed, damaged, or stressed in the current conflict. The $13.6 billion in reset funds in the FY 2008 GWOT Request for the U.S. Army will go a long way towards raising the readiness levels across the force.

## Force Protection

This FY 2008 GWOT Request includes $15.2 billion for investments in new technologies to better protect our troops from an agile and adaptive enemy. Programs being funded would include a new generation of body armor, vehicles that can better withstand explosions from Improvised Explosive Devises (IEDs), and electronic devices that interrupt the enemy's ability to attack U.S. forces. Within this force-protection category, the FY 2008 GWOT Request includes $4 billion to counter and defeat the threat posed by IEDs.

## Afghan/Iraqi Security Forces

The FY 2008 GWOT Request includes $4.7 billion to stand up capable military and police forces in Afghanistan and Iraq. The bulk of these funds are going to train and equip Afghan National Security Forces (ANSF) to assume the lead in operations throughout Afghanistan. As of January, 88,000 have been trained and equipped, an increase of 31,000 from the previous year.

In Iraq, more than 300,000 soldiers and police have been trained and equipped, and are in charge of more than 60 percent of Iraqi territory and more than 65 percent of that country's population. They have assumed full security responsibility for three out of Iraq's 18 provinces and are scheduled to take over more territory over the course of the year. These Iraqi troops, though far from perfect, have shown that they can perform with distinction when properly led and supported. Iraqi forces will be in the lead during operations to secure Baghdad's violent neighborhoods. As we significantly increase and improve the embedding program, Iraqi forces will operate with more and better Coalition support than they had in the past.

## Non-Military Assistance

Success in the kinds of conflicts our military finds itself in today—in Iraq, or elsewhere—cannot be achieved by military means alone. The President's strategy for Iraq hinges on key programs and additional resources to improve local governance, delivery of public services, and quality of life—to get angry young men off the street and into jobs where they will be less susceptible to the appeals of insurgents or militia groups.

Commanders Emergency Response Program, or CERP funds, are a relatively small piece of the war-related budgets—$977 million in the FY 2008 GWOT Request. But because they can be dispensed quickly and applied directly to local needs, they have had a tremendous impact—far beyond the dollar value—on the ability of our troops to succeed in Iraq and Afghanistan. By building trust and confidence in Coalition forces, these CERP projects increase the flow of intelligence to commanders in the field and help turn local Iraqis and Afghans against insurgents and terrorists.

## Conclusion

With the assistance and the counsel of Congress, I believe we have the opportunity to do right by our troops and the sacrifices that they and their families have made these past few years. That means we must make the difficult choices and commit the necessary resources to not only prevail in the current conflicts in which they are engaged, but to be prepared to take on the threats that they, their children, and our nation may face in the future.

## Army Chief of Staff Change of Responsibility Ceremony

*Fort Myer, Virginia, Tuesday, April 10, 2007*

THANKS, Pete.
The only positive thing I can find about Pete Schoomaker retiring is that it will simplify my life a little bit as I deal with Pete Geren, Pete Pace, Pete Chiarelli.

I want to thank all of our distinguished guests, especially the former chiefs of staff of the Army who are here with us today.

I've been fortunate in recent weeks to attend a number of changes of command, and they've given me the opportunity to honor some of our military's most talented leaders—some of whom are going on to other commands, and some of whom are retiring after decades of service to our country.

General Schoomaker falls into the latter category—but I would have to note this isn't the first time he retired. I do suspect it will be the last.

Some of his staff were under the impression that his truck is over in the parking lot right now, packed, with the engine running as we speak—just waiting for this ceremony to end. Rumor has it that Pete has a vanity license plate that reads "AWOL."

As Pete Geren noted, in 2003, Secretary Rumsfeld asked Pete to return to service as Army Chief of Staff, and it's a great testament to Pete's sense of duty that he abandoned a nascent ranching venture in his home state of Wyoming to return to Washington, D.C.—a place where, he has noted, too often the horses ride the cowboys.

Like Pete, I flunked from retirement, but unlike Pete, who as the other Pete said, hung up on Secretary Rumsfeld thinking it was a prank caller, I did not hang up on the President—a decision I may live to regret.

But it does remind me of another Washington saying. You know about being careful what you ask for—well, in Washington you have to be careful even what you don't ask for, because you may still get it.

Although Pete didn't ask for this post, he knew from the second he took the oath that there were great challenges before him, that this was a critical moment in the history of the United States and in the history of the Army.

I doubt many people know it, but both Pete and I consider the same event to have been a pivotal moment in our professional lives. As Secretary Geren said, in early 1980, Pete was part of the team sent in to rescue our hostages from the embassy in Iran. I was chief of staff at the CIA at the time, and I'd spent most of that long night at the White House with the director of Central Intelligence.

Though we at the White House and those on the ground saw things from very different perspectives, many of us came away with the same lessons about the importance of true jointness and the constant need to be prepared and vigilant; to maintain our armed forces even in times of peace, and always to look ahead to threats on, and even beyond, the horizon.

Pete keeps a photo of the destruction from the rescue attempt to remind him of what had happened, and it was at that moment that he committed himself to a future where enthusiasm would always be matched by our capability.

We are seeing that future today. Challenging times require extraordinary vision and leadership, and Pete has shown both of those qualities. He has entirely changed the manner in which our Army is trained, deployed and organized.

Pete was called out of retirement because of his wealth of experience, particularly his unique knowledge of special operations and unconventional warfare—an area of expertise that stresses innovation and versatility. He has shown remarkable ability to lead individuals as well as institutions during his more than 30 years in the military—from his days as a platoon leader, to several Delta Force assignments, to his leadership of U.S. Special Operations Command.

Preparing our forces for the kinds of wars we are fighting—and the ones we may be called upon to fight in the future—is a difficult task in an environment that requires the riflemen as well as the "smart bombs," unconventional approaches as well as conventional power.

Pete has led the transition from a division-based Army—the standard since World War II—to a brigade-based Army, a lighter more lethal force that can deploy rapidly and effectively to meet today's challenges. To give some perspective, when I was last in government 14 years ago, we measured the time it took to deploy most brigades in months. Today, we measure it in weeks and days.

In just a few short years, Pete has also revamped the training protocol across the entire Army, focusing on the skills necessary to conduct complex operations on an unconventional battlefield with no "frontlines."

I'm reminded of a story Stephen Ambrose told in one of his books about World War II. A reporter was interviewing a man who had been a in front-line foxhole during the Battle of the Bulge, and he asked about the "rear echelon."

And the veteran replied, "Listen, as far as I'm concerned, every son of a bitch behind my foxhole is rear echelon." That sounds a lot like Pete Schoomaker.

Every so often, an institution needs a leader to remind us of its core values. Pete has done that by emphasizing the "warrior ethos" and focusing on physical fitness and basic skills like marksmanship and hand-to-hand combat. This led to a renewal of timeless values, like personal courage and pride in one's physical and mental strength—integral parts of the moral fiber and institutional memory that has, throughout history, made our military so effective against our enemies, and so respected by our friends.

Any of these accomplishments alone would be the accomplishment of a career—but Pete has managed to do all of this in a few short years while simultaneously fighting two wars.

And he has done so within the confines of one of the world's largest bureaucracies—one that isn't exactly known for turning on a dime. In fact, I understand that when Pete came out of retirement, his status was changed somehow from "retired" to "deceased." It took General Schoomaker, the highest-ranking officer in the Army, a full six months to iron out the paperwork. Leave it to the Pentagon bureaucracy to prove that you can in fact be brought back from the dead.

Pete, thank you for your service. And, Cindy, he couldn't have done it without your love and support. Our nation is grateful to both of you—and the men and women of the armed forces are stronger and safer, because of everything you've done.

Pete's successor, General George Casey, is well-qualified to take the Army's spot—or was qualified—well-qualified to take the Army's spot three years ago after serving as Vice Chief.

After that job, however, he volunteered to spend 30 months as commander of Multi National Forces-Iraq.

In that capacity, he oversaw the largest sustained ground-force operation by the United States military in more than 30 years—a tenure that included the ratification of the Iraqi constitution, two successful nationwide elections, and the creation of the Iraqi army and police force essentially from scratch.

George Casey has the unique experience of having served at the highest levels on both the institutional and operational sides of the Army. Perhaps more importantly, though, he has seen the face of war in the 21st century firsthand—the complex nature of asymmetric warfare, urban combat, counterinsurgency operations, and sustained commitments of a rotational, expeditionary Army abroad.

If George Casey were well-qualified to take this position before his tour in Iraq, he is superbly qualified now.

He has an eminent familiarity with all the challenges—personnel, equipment, and tactics—that the Army must face in the present and in the future.

I'd be remiss if I didn't close without a word—with a word about Sheila Casey. For 30 months, she endured separation from George when he agreed and then re-agreed to assume the mantle and burdens in command in Iraq. Our nation is in Sheila's debt, as we are to all the spouses of all the soldiers who are deployed to dangerous and distant battlefields. I know Sheila will be a stalwart advocate for Army families as she takes on her latest role in what has been a lifetime of service.

George, Sheila, thank you for everything you have done and will do for this country. I look forward to working with you.

Thank you very much.

## American Chamber of Commerce of Cairo

*Cairo, Egypt, Wednesday, April 18, 2007*

Thank you for that kind introduction, and my appreciation to the American Chamber of Commerce for the invitation to speak today.

This afternoon, I'd like to discuss the U.S.-Egypt relationship and the security challenges facing us in the Gulf region. These challenges, though significant, can be overcome with leadership and commitment from both of our nations.

I have long considered Egypt one of America's most important, even indispensable, partners. During my earlier years in government—both at CIA, and at National Security Council—I had a chance to witness the relationship between our two countries being forged after years of animosity. I was here in the late 70s for meetings with President Sadat during the final stages of the Camp David Accords, here again in the mid 80s as Deputy Director of Central Intelligence, and again in 1990 to discuss with President Mubarak our joint efforts in Operation Desert Storm to liberate Kuwait.

The development of our friendship in the 1970s proved to be one of the bright spots in a decade that was perceived to be a very difficult period for the United States in the context of the Cold War—with setbacks in southeast Asia, Latin America and, of course, with the hostages in Iran. It was a relationship that began with shuttle diplomacy and peace negotiations with Israel, and has since evolved into a strong bilateral partnership of its own.

Since 1980, the military relationship between the U.S. and Egypt has been expressed through the Bright Star exercise—a joint training event that now includes forces from multiple nations. This joint venture is just one example of our commitment to building the capabilities of the Egyptian armed forces. In addition to the more than $1 billion in military aid that Egypt will receive from the U.S. this year, we continue to maintain and strengthen the ties between our military establishments through education, training, and exchanges.

Our own military—in particular the American officer corps—has certainly benefited and learned a good deal from working with the Egyptian military, which is one of the region's most professional and effective forces.

Some of the most consequential progress in our relationship in recent years has not come between our militaries, or even between our governments, but as a result of the work of many of you here today. The growth of trade has brought Egypt closer to the global marketplace of investment, commerce, and ideas—and that is a welcome development.

Since Ronald Reagan's administration, American presidents have benefited from the counsel of President Mubarak, whom I first met during that visit in August 1990. The United States has welcomed the role he has played as a key broker between Israeli and Palestinian leaders in recent years, through the Roadmap, and as a facilitator in the overall regional peace process.

In my meetings with senior Egyptian officials today, I reaffirmed our important military, diplomatic, and economic ties. Together, we reiterated our shared goals on some of the most pressing issues in this region and of our time. They include:

- A unified, stable, and prosperous Iraq;
- A just and comprehensive peace between the Israeli and Palestinian people;
- An Iran that does not attempt to dominate the region by subverting their neighbors and by building nuclear weapons; and
- Halting the growth and influence of extremist terrorist networks and sectarian militia organizations that have become, in the words of our former theater commander, "the curse of the region."

The issue of Iraq continues to dominate the political landscape of the United States, and is uppermost in the minds of the people of the Middle East who have watched developments in that country with growing concern. In recent months, the United States has reaffirmed its commitment to Iraq, and with that, our commitment to protect our allies and long-standing interests in this region. A new military strategy is in its initial phases—a strategy focused on providing basic security to the Iraqi people. It is being bolstered by a new emphasis in the political, economic, and governance areas designed to improve the quality of life for all Iraqis.

The immediate goal is to create the breathing room necessary to allow reform and reconciliation to go forward—steps that will give all of Iraq's communities, majority and minorities alike, a stake in that nation's future.

As I've said to many of our long-standing friends in recent months, whatever disagreements we might have over how we got to this point in Iraq, the consequences of a failed state in Iraq—of chaos there—will adversely impact the security and prosperity of every nation in the Middle East and Gulf region. There may be some, who, over resentment or disagreements over what happened in the past, might be cheering for failure.

I would respectfully suggest that these sentiments are dangerously short-sighted and self-destructive. The first and secondary effects of a collapse in

Iraq—with all of its economic, religious, security and geopolitical implications—will be felt in capitals and communities of the Middle East well before they are felt in Washington or New York. The forces that would be unleashed—of sectarian strife, of an emboldened extremist movement with access to sanctuaries—do not recognize or respect national boundaries.

For this reason, Iraq's neighbors will need to play a constructive role going forward. The regional talks recently held in Baghdad with Egypt's support were a good start toward improved cooperation, and our government is open to higher-level exchanges.

We welcome the following meeting in Sharm al-Sheikh and appreciate President Mubarak's constructive role in organizing it.

We encourage Iraq's Arab neighbors to use their influence to dampen homegrown insurgency and alleviate sectarian conflict. Other nations, who have not been "good neighbors" to Iraq—such as Syria and Iran—should start becoming part of the regional solution that encourages political reconciliation and reduces violence.

In the case of Iran, it is no secret that, as a private citizen, I advocated for dialogue on some issues. But as we were reminded recently with regard to the British sailors, in dealing with a regime like Iran's, one has to be realistic.

I remember back to November 1, 1979, when then-National Security Advisor Zbigniew Brzezinski and I met with the Iranian political leadership in Algiers. We discussed the possibility of continuing the partnership that had previously existed under the Shah—including military assistance to the new government. Three days later came word that 66 of our diplomats had been seized in Tehran, and two weeks later, the prime minister and defense and foreign ministers with whom we had met were out of their jobs.

We should have no illusions about the nature of this regime—or about their designs for their nuclear program, their intentions for Iraq, or their ambitions in the Gulf region.

There is also the threat posed by terrorist networks and their financial and ideological sponsors—a threat that transcends nations and continents.

It is important to remember that in Iraq, the primary victims of Al Qaeda and their affiliates have not been Coalition troops or Iraqi security forces, but tens of thousands of innocent civilians—men, women, and children whose major crime was to go to the market or attend Friday prayers. Where the extremists have seized and controlled territory in the past—in western Iraq, eastern Afghanistan, or elsewhere—the result has been misery, poverty, and fear. We have seen the future promised by the extremists: a dark, joyless existence personified not by piety or virtue, but by the executioner and the suicide bomber.

To overcome these daunting challenges—defeating the terrorist networks, securing Iraq, holding Iran accountable, bringing peace to the Holy Land—geography and history have thrust an important and unique burden upon Egypt. It is a role well in keeping with Egypt's historic tradition of providing leadership in the Arab world.

In fact, I would note that many of the most important developments in this region have begun with Egypt. It was this way during the Cold War—in forming an alliance with the former Soviet Union, and then expelling it; in fighting five wars with Israel, and then being the first to make peace. When Egypt has had the courage and vision to lead—despite the real risks and costs involved—it has benefited not only Egypt, but the people of the Middle East as well.

Because of Egypt's unique position—its geography, economy, and demographics—it is unlikely that progress can be made on the most pressing issues of today without Egypt's full engagement, support, and leadership. But with Egypt providing leadership, it will be possible to open up new possibilities for the people of the Middle East. After all, we are living at a time where, as never before, people around the globe are demanding and making progress toward peace, political openness, and an economic system that works for themselves and their families.

For all the difficulties we now confront—and they are daunting, to be sure—I believe that, over time, there can be a very different future—a different narrative, if you will—for this part of the world. A future:

- Where trade, commerce, and economic opportunity lead to a growing middle class and a higher quality of life for workers and their families;
- Where Palestine and Israel are living in peace side by side as viable and independent states;
- Where men and women have an increasingly greater say and a greater stake in how they govern their own lives, their own communities, and their own countries; and
- Where citizens from Tehran to Baghdad to Beirut can look forward to a life secure from the assassin, the suicide bomber, and the proverbial knock on the door in the middle of the night.

Reaching these goals cannot be achieved by any one nation alone—no matter how wealthy or powerful. And they certainly cannot be achieved solely by military means. To do all this, we all—the United States, Egypt, and other key players in the region—must be engaged. And we must lead. And we must work together.

You may read and hear of heated debates in the United States over the course our nation should take in Iraq. I suspect that these debates reflect discussions taking place in coffee shops and in conference rooms across this country and the Middle East generally. Friends and adversaries alike are watching America closely and may have questions about our commitments and intentions—in this region, and around the world.

I am here today to reaffirm what multiple administrations of both American political parties have concluded: that the relationship between the United States and Egypt is vital and enduring, and that our own security and prosperity is closely linked to the security and prosperity of this part of the world. To build a more secure and prosperous future we will continue working with Egypt and other friends in the region: not as a patron, but as a partner—a partner that respects the different histories, cultures, and perspectives of the people of the Middle East. It is a responsibility we will not abandon, a trust we will not break.

Once again, I thank the chamber for the opportunity to spend time with you today, and I look forward to your questions.

# Navy Flag Officers Conference

*U.S. Naval Academy, Wednesday, April 25, 2007*

GOOD afternoon.

It's a pleasure to have this opportunity to talk to you, and to give you a chance to ask a few questions.

I've accepted an invitation to speak at the Naval Academy's commencement in a few weeks—and I'm really looking forward to meeting firsthand the midshipmen who will soon be joining your ranks. I presided over 39 commencements as president of Texas A&M, but this will be my first commencement speech ever. Only thing I know for sure is to keep it short.

The character of any institution is determined by the character of its individuals—and I know the graduating class will make a great contribution to the Navy.

I'd like to take a few minutes and talk about what I see as some of the most important priorities and goals for the Navy in the years to come. Some of these priorities relate to the threats of today, and others relate to ensuring that our country remains vigilant and prepared to meet the threats of tomorrow.

First of all, let me say that I appreciate the Navy's direct and indirect support of operations in Iraq, Afghanistan, and elsewhere in the War on Terror. The nature of the fight is such that the Department has asked sailors to take on duties that they would not ordinarily do—such as building schools and hospitals or helping provide security for convoys. Considering the very different types of operations called for by conflicts like those we are engaged in, it makes sense to consider the Navy and the Air Force a "Strategic Reserve."

At the same time, however, we should not forget that even in times of peace, the Navy is much more than just a Reserve force. Our fleet is a symbol of America's presence across the globe—a deterrent to all of our enemies, and a reassurance to all of our allies that we will keep our commitments. It was in this spirit that I sent a second carrier strike group to the Gulf region.

With the types of globalized and uncertain threats we face today, we have to focus on the Navy's unique war-fighting capabilities. For this reason, I support the chief of the Navy's maritime strategy, which better aligns budgetary decisions with our risk assessments and future missions we may have to undertake—and which also underscores the interdependence of the Navy-Marine Corps team.

We should not forget that in this age no single nation is capable of addressing the myriad threats we face. This is why I strongly support the 1,000-ship Navy initiative. Any effort to bring together the navies of multiple nations in either a formal or informal manner is well worth the effort. Not only is this in line with the President's and the Quadrennial Defense Review's call for partner-building, but it is also the only way we can meet threats that do are not limited to any single nation or region.

On a more operational level, I agree with the chief that our three priorities must be readiness, building a fleet for the future, and developing leaders for the 21st century.

To ensure readiness, we have to acknowledge that readiness is something that must be maintained over time. It is not a sprint to any sort of finish line. This means we have to continue focusing on initiatives that increase our long-term readiness—initiatives like the Fleet Response Plan, which has already greatly bolstered our ability to surge the fleet if the need arises—while still meeting all the needs of the combatant commanders. Like everything to do with readiness, the Fleet Response Plan is an ongoing project, but one with even greater potential in the future.

Long-term readiness is also to a large degree contingent on building a fleet for the future. The nature of the defense industry is such that it requires long-term thinking and planning—and it requires a careful evaluation and management of risks as we consider what types of ships we may need many years from now. We have to be flexible in our thinking—but at the same time we have to provide some sort of stability for the defense industry, so they can make appropriate business decisions. We need to move ahead with the current plan for 313 ships and around 3,800 aircraft since at this time this appears to be the minimum force we will need in the future.

None of this is possible without effective leaders. As a former university president, I applaud the chief's drive to develop an education strategy for the Navy, especially one that places so much emphasis on the skill that will be important in the future—skills that reflect the evolution of technology, the interdependence of the branches, and the sophisticated regional knowledge that is so crucial in a globalized world. Young people today have many opportunities, so we will have to always be on the lookout for better ways to compete for the most talented students—whether that means taking cues from the corporate world or increasing our minority recruitment.

I would just close with a few words about intelligence and its importance today. Having served most of my career at CIA, no one appreciates the role intelligence plays more than I do. I have, however, seen control over intel divide organizations or slow them down with internal fights. In today's world, we need to put any feeling like that aside. With the types of enemies we face, intel has become more important than ever—and we have also learned that while intel can be a dividing force, it can also bring us together and increase cooperation, both domestically and internationally.

We face many challenges today, and we will also face many more in the years ahead. But if we can maintain a spirit of cooperation—within the branches, among the branches, domestically, and internationally—then I have no doubt that our nation will continue to be the leader of the free world for decades to come.

I'll close with a few words about two institutions of importance to us—the Congress and the press. Too often we see both as adversaries. They are not. The Congress is a co-equal branch of government that under the Constitution raises and supports armies and navies. Members of both parties now in Congress have been strong supporters of the Department of Defense, especially our men and women in uniform, for a long time. As senior officers and as civil servants, you have a responsibility to communicate to those below you that the American military must be non-political and realize the obligation we owe the Congress to be honest and true in our reporting to them, even when it means admitting mistakes or problems.

The same is true with the press, in my view a critically important guarantee of our freedom. When they identify a problem—as at Walter Reed—the response of senior leaders should be to find out if the allegations are true, as they were at Walter Reed. And if so, to say so, and then act to remedy the problem. If untrue, then be able to document that fact. But they are not the enemy, and to treat them as such is self-destructive.

As the founding fathers wisely understood, the Congress and a free press, as with a non-political military, assures a free country.

So with that, I'll be happy to take a few questions, if there's anbyody among you courageious enough to start the process.

## Greater Dallas Chamber of Commerce

*Dallas, Texas, Thursday, May 03, 2007*

THANK you for that kind introduction, Erle and a special thanks to the Greater Dallas Chamber of Commerce for the invitation to speak today.

First things first: Howdy!

You ought to try starting a congressional hearing sometime with that. It is a nice way to begin a speech. In Washington, most of my public remarks tend to begin with someone asking me to raise my right hand. And then asking if I'm actually going to tell the truth. Well, it's a pleasure to be here in Dallas, and to be back in Texas. Of course, it's a pleasure to be anywhere but Washington, D.C. Where those who travel the high road of humility encounter little heavy traffic. The only place in the world you can see a prominent person walking down lover's lane holding his own hand. Sometimes back there I'm reminded of President Truman's comment about his arrival in Washington. He said, "For the first six months, you wonder how the hell you got here. For the next six months, you wonder how the hell the rest of them ever got here."

It is a pleasure to be back in Texas, where, as Erle mentioned, I spent four and a half wonderful years at Texas A&M, and I understand from the applause earlier that there is a healthy contingent of Aggies today. There is a certain symmetry. I was the—as Earl mentioned—I'm the 22nd Secretary of Defense. I also happened to be the 22nd President of Texas A&M. You'll notice that Texas A&M was founded in 1876, and has had 22 presidents and the Department of Defense was founded in 1947, and has had 22. There's a message there somewhere.

The Aggies will be pleased to know that the senior boots given to me by the Corps of Cadets are on proud display in the office of the Secretary of Defense. Longhorns are not required to genuflect. And their career prospects have not been damaged ... much.

To the members of the Greater Dallas Chamber, let me express my deepest gratitude to all of you who are supporting your Guard and Reserve employees while they are mobilized. You value these citizen soldiers and know they're more than worth retaining. Thank you for believing in them.

This is a tough time for them and for their families—and I can assure you that your help and understanding do not go unnoticed.

There are a lot of other activities your members sponsor that benefit the military, from Adopt-a-Soldier projects, to providing complimentary sports tickets, to helping service men and women travel to be with their loved ones or keep in touch while overseas.

And of course, the Dallas-Fort Worth airport has its "Welcome Home a Hero" initiative—a great program that organizes local groups, schools, congregations, and businesses to be on hand every single day to greet men and women returning from their deployments. This morning, I met a flight from Kuwait into Dallas-Fort Worth, with almost 200 service members from several branches of the armed forces.

These kinds of public receptions are really important. Whatever disagreements exist over the war in Iraq, we are all united in our admiration of the men and women who have volunteered to serve our nation during these challenging times. These heartwarming receptions are a far cry from what took place in the late 1960s and the 1970s, and it's good to see that we have learned from our past experiences and mistakes.

Nowadays a lot of people do want the troops to know they care, yet some aren't sure how to get involved. A few years ago, the Department of Defense launched a program called "America Supports You" that recognizes all the things that are being done across the country to show appreciation for the U.S. military and their families. If people go to the "America Supports You" website, they will see countless projects they can join—from sending care packages to troops abroad, to lending a helping hand to their families here at home. Everything you do makes a difference.

In the relatively short time since I became Secretary of Defense, I've had the opportunity to meet and work with some extraordinary people dedicated to serving their country. At the top of the list are our troops on the front lines in Iraq and Afghanistan. I've come away from my four visits in four months to Iraq and Afghanistan impressed by their resilience, their good humor, their courage, and their determination in the face of danger and personal sacrifice.

And I've been just as impressed with the wounded troops at Walter Reed. To be honest, before I went the first time, I dreaded it. But people kept telling me, "No, you don't understand, they'll lift you up."

And they did. And they do whenever I visit there. Especially the wounded officer at Walter Reed who reminded me that I had handed him his diploma at Texas A&M in August of 2002. He also told me he had the doctors play the "Aggie War Hymn" during his surgery. That's commitment. All of these young men and women are so impressive, and the military and the federal government are putting a full-court press on to make sure we solve the problems relating to outpatient care that have come to light in recent weeks.

Not all parts of the job are so edifying. This is budget season in Washington—another reason I looked forward to being here today. One of my first duties when I became Secretary was to present the Department's base budget and war requests for the next fiscal year. I'm here to tell you you haven't lived until you've gone before Congress and asked for nearly three quarters of a trillion dollars from the taxpayers' wallets. It gives a whole new meaning to the term "sticker shock."

But I also believe that it's important to put that defense budget in some historical context. Consider that, at about four percent of America's gross domestic product, the amount of money the United States expects to spend on defense this year is actually slightly a smaller percentage of GDP than when I left the government 14 years ago, following the end of the Cold War—and a significantly smaller percentage of GDP than during previous times of war, such as Vietnam and Korea.

Someone once said that "experience is that marvelous thing that enables you to recognize a mistake when you make it again."

Five times over the past 90 years—after the First and Second World Wars, Korea, Vietnam, and most recently, after the Cold War—the United States has slashed defense spending or disarmed outright in the mistaken belief that the nature of man or the behavior of nations had changed, with the end of each of the wars. Or that somehow we would not face threats to our homeland or would need to take a leadership role abroad. Win, lose, or draw, we cut back at those times, not just in defense, but in other vital elements of national power, such as diplomacy and intelligence.

In 1992, when I was Director of Central Intelligence, I testified before Congress when it was considering CIA's budget at the end of the Cold War. I noted that while we had just witnessed the implosion of the Soviet empire, the end of the superpower struggle by no means meant that history was over. Rather, I argued that history had been temporarily frozen by the Cold War and by World War II, and might well thaw with a vengeance. Intelligence officers are traditionally pessimists. But in the early 1990s, not even I would have predicted how rough history's thaw would be.

There's a lot that's good about American optimism. But on occasion that optimism has led us to believe that if we would just deal with this or that crisis abroad, we could then pay less heed to the rest of the world, and then turn inward and pay attention to domestic affairs. That's why our nation, once again, pursued a false "peace dividend" in the 1990s. With the complicity of both political parties, key instruments of America's national power—military, diplomatic, and intelligence—withered throughout the decade.

Consider that between 1989 and 2001, the active Army declined from almost 800,000 troops to fewer than 500,000. The number of Navy ships decreased by more than half. And the Air Force went from 37 tactical wings to 20.

As CIA Director, I tried to fight off reductions in the intelligence budget. While CIA's spending and manpower figures are closely held, the public knows through the findings of the 9/11 Commission that by the mid-1990s, recruitment of new case officers at CIA had hit a historic low, and the Agency's funding was a prime target for budget-cutters. Indeed, within three years of my retirement in 1993, CIA's clandestine service had been cut by 30 percent—just when Usama bin Laden was gearing up his war on the United States.

The place where we may have taken the biggest hit was in our ability to engage, assist, and communicate with other parts of the world. The State Department froze new hiring of foreign service officers. The Agency for International Development saw deep staff cuts and relied more and more on outsourcing—its permanent staff dropping from a high of 15,000 during the Vietnam era to 3,000 today. In what had been an enormously successful organization for communicating America's values and message abroad during the Cold War—the United States Information Agency—was abolished in the 1990s as an independent entity and folded into the State Department.

But even as we throttled back, the world became more unstable, more turbulent, and more unpredictable than the Cold War years we had left behind. Our hopes for peace, once again, gave way to the realities of disorder and conflict.

The attacks of September 11, 2001, abruptly ended our "holiday from history," and the illusions that came with it. Those attacks, and the subsequent campaigns in Afghanistan, Iraq, and around the globe, revealed military, intelligence, and diplomatic shortcomings that our government has had to work hard to correct.

As a nation, I believe we have to do two things. First, we have to deal with the challenges we face now—to commit the necessary resources to the current campaigns in Iraq and Afghanistan. And second, we must recoup the underinvestment of the past, and commit ourselves to strengthening the instruments of national power across the board.

This won't be easy. As General Pete Schoomaker liked to say about some of his efforts as Chief of Staff of the Army, this process is like tuning a car engine while the vehicle is moving.

Most pressing to me as Secretary of Defense has been the stress the ongoing conflicts has placed on our armed services, and in particular our ground forces: the Army and the Marine Corps, who bore the brunt of under-funding in the past and the bulk of the costs—both human and material—of the wars of the present. In fact, investment in Army equipment and other essentials was underfunded by more than $50 billion before we invaded Iraq.

In addition to allocating tens of billions of dollars to repair or replace damaged or destroyed equipment, we will also be increasing the size of the active Army and Marine Corps by some 92,000 over the next five years. One of the effects of this increase, over time, should be that with a larger pool of ground forces, it will be less necessary to call on National Guard units as often for deployments overseas. We are also recapitalizing and modernizing our air, sea, and land forces for all kinds of conflicts—whether they are conventional battles or counterinsurgency campaigns—that may come in the future.

A rethinking of strategy and redirection of resources is taking place throughout the government. I note in this connection the change in our diplomatic posture being directed by the Secretary of State, Condi Rice. As the Secretary noted last year, it makes little sense that our embassy in Germany, a nation with a population of 82 million, has about the same number of U.S. State Department officers as our embassy in India, with over a billion people. The State Department has begun to shift its assignments and post locations to better suit the realities and priorities of the 21st century.

In the area of intelligence, I am encouraged that CIA has dramatically ramped up recruitment and has boosted our human intelligence efforts.

The goal is an integrated effort, a reinvigoration of the key elements of national power so that the United States does not let down its guard again. We will not achieve the goal tomorrow or the day after tomorrow. But I think we have at least made a start.

This is a crucial time for our country, our nation—for all our allies. Violent extremist networks and ideologies will continue be a threat to the United States and our allies for many years. The ambition of these networks to acquire chemical, biological, and nuclear weapons is real, as is their desire to launch more attacks on our country and on our interests around the world.

There are other areas of concern:
- The nuclear ambitions and missile programs of Iran and North Korea, which pose a threat to their neighbors and the entire world because of their record of proliferation;
- The uncertain paths of Russia and China, which are both pursuing sophisticated military modernization programs;
- And a range of other potential flashpoints.

There is no doubt that the challenges we face are daunting. And certainly it can be depressing to read the latest news bulletin from the Middle East. But I also remember what things were like when I first went to Washington in the summer of 1966, at the height of the U.S. buildup in Vietnam.

What lay ahead were:
- Violent domestic turmoil;
- Two assassinations at home of historic consequence;
- A major war in the Middle East;
- The seizure of a U.S. Navy ship by North Korea;
- The Soviet invasion of Czechoslovakia;
- And the resignation of a president in disgrace;

And by the end of the 1970s we saw:
- A collapse in Vietnam, and the deaths of millions across Southeast Asia;
- High inflation;
- High interest rates;
- Two energy crises;
- The Soviet invasion of Afghanistan;
- Revolution in Iran and an embassy taken hostage;
- And tens of thousands of Cuban soldiers in Angola and Ethiopia;

By 1980, the Soviet Union seemed ascendant and we were reeling.

Who would have anticipated during that discouraging period, the groundwork was being laid—through policies pursued by administrations of both political parties—for the remarkable turn of events that occurred a decade later: the fall of the Berlin Wall, the liberation of Eastern Europe, victory in the Cold War, the reunification of Germany, the dissolution of the Soviet Union, the liberation of hundreds of millions of people behind the Iron Curtain and around the world, and an American economy beginning more than a quarter of a century of vibrant economic growth.

There are, I think, two lessons from all this. First, our weariness with conflict—with the setbacks and tragedy of war—is understandable and even to be expected. But second, we must not let that weariness cause us to withdraw from the world or diminish our ability to deal with the threats and challenges of tomorrow. There is no way to predict the future, nor can we predict the effect that decisions made today will have a decade or two from now.

One thing is clear, though, from history: when America is willing to lead the way; when we meet our commitments and stand with our allies, even in troubling times; when we prepare for threats that are on the horizon and beyond the horizon; and when we make the necessary sacrifices and take the necessary risks to defend our values and our interests—then great things are possible, and even probable for our country and the world.

A final thought: our country is troubled and divided by a long and difficult war in Iraq. We want our troops to come home and be out of harm's way. And yet, most know—or at least sense—that leaving chaos behind us in Iraq will bring dramatically more suffering for Iraqis and also disaster for the Middle East—and ultimately for us.

At such a difficult time, it is perhaps fitting to close with the words of Winston Churchill, who said of us, "The price of greatness is responsibility ... The people of the United States cannot escape world responsibility." That was true when Churchill said it in 1943, and it is still true nearly 65 years later.

As a nation, we have over more than two centuries made our share of mistakes—from time to time strayed from our values, and on occasion become arrogant in our dealings with others. But we have always corrected our course. And that is why today, as throughout our history, this country remains the world's most powerful force for good. Because we stand for liberty and we stand for the God-given worth of each and every person. This country will continue to be a beacon for all who are oppressed. And it will continue to accept its responsibility for leadership in the world. And that is good news for the world.

Thank you.

## Senate Appropriations Committee

*106 Dirksen Senate Office Building, Washington, DC, Wednesday, May 09, 2007*

MR. Chairman, members of the Committee:
I thank the Committee for all you have done to support our military these many years, and I appreciate the opportunity to provide an overview of the way ahead at the Department of Defense through the President's Fiscal Year 2008 Defense Budget, which includes the base budget request and the FY 2008 Global War on Terror Request.

I believe that it is important to consider the budget requests submitted to the Congress this year—the base budget and the war-related requests—in some historical context, as there has been, understandably, sticker shock at their combined price tags—more than $700 billion total.

But consider that, at about 4 percent of America's gross domestic product, the amount of money the United States is expected to spend on defense this year is actually a smaller percentage of GDP than when I left government 14 years ago, following the end of the Cold War—and a significantly smaller percentage than during previous times of war, such as Vietnam and Korea.

Since 1993, with a defense budget that is a smaller relative share of our national wealth, the world has gotten more complicated, and arguably more dangerous. In addition to fighting the Global War on Terror, we also face:

- The danger posed by Iran's and North Korea's nuclear ambitions and missile programs, and the threat they pose not only to their neighbors, but globally, because of their records of proliferation;
- The uncertain paths of China and Russia, which are both pursuing sophisticated military modernization programs; and
- A range of other potential flashpoints and challenges.
- In this strategic environment, the resources we devote to defense should be at the level to adequately meet those challenges.

Five times over the past 90 years the United States has either slashed defense spending or disarmed outright in the mistaken belief that the nature of man or behavior of nations had somehow changed, or that we would no longer need capable, well funded military forces on hand to confront threats to our nation's interests and security. Each time we have paid a price.

The costs of defending our nation are high. The only thing costlier, ultimately, would be to fail to commit the resources necessary to defend our interests around the world, and to fail to prepare for the inevitable threats of the future.

As Sun Tzu said more than 2,500 years ago, "The art of war teaches us to rely not on the likelihood of the enemy's not coming, but on our own readiness to receive him; not on the chance of his not attacking, but rather on the fact that we have made our position unassailable."

A perspective in this regard—closer in time and place to today—is that of George Washington who said in his first [State of the Union] address, "To be prepared for war is one of the most effectual means of preserving peace."

## FY 2008 BASE BUDGET

The President's FY 2008 base budget request of $481.4 billion is an increase of 11.4 percent over the enacted level of FY 2007, and provides the resources needed to man, organize, train, and equip the Armed Forces of the United States. This budget continues efforts to reform and transform our military establishment to be more agile, adaptive, and expeditionary to deal with a range of both conventional and irregular threats.

Some military leaders have argued that while our forces can support current operations in the War on Terror, these operations are increasing risks associated with being called on to undertake a major conventional conflict elsewhere around the world. This budget provides additional resources to mitigate those risks.

The FY 2008 base budget includes increases of about $16.8 billion over last year for investments in additional training, equipment repair and replacement, and intelligence and support. It provides increases in combat training rotations, sustains air crew training, and increases ship steaming days.

## INCREASE GROUND FORCES

Despite significant improvements in the way our military is organized and operated, the ongoing conflicts in Iraq and Afghanistan have put stress on parts of our nation's ground forces.

In January, the President called for an increase in the permanent active end strength of the Army and Marine Corps of some 92,000 troops by FY 2012. The base budget request adds $12.1 billion to increase ground forces in the next fiscal year, which will consist of 7,000 additional Soldiers and 5,000 additional Marines.

Special Operations Forces, who have come to play an essential and unique role in operations against terrorist networks, will also grow by 5,575 troops between FY 2007 and FY 2008.

## Strategic Investments—Modernization

The base budget invests $177 billion in procurement and research and development that includes major investments in the next generation of technologies. The major weapons systems include:
- Future Combat System ($3.7 billion)—The first comprehensive modernization program for the Army in a generation.
- Joint Strike Fighter ($6.1 billion)—This next generation strike aircraft has variants for the Air Force, the Navy, and the Marine Corps. Eight international partners are contributing to the JSF's development and production.
- F-22A ($4.6 billion)—Twenty additional aircraft will be procured in FY 2008.
- Shipbuilding ($14.4 billion)—The increase of $3.2 billion over last year is primarily for the next generation aircraft carrier, the CVN-21, and the LPD-17 amphibious transport ship. The long-term goal is a 313-ship Navy by 2020.

## Missile Defense

I have believed since the Reagan administration that if we can develop a missile defense capability, it would be a mistake for us not to do so. There are many countries that either have or are developing ballistic missiles, and there are at least two or three others—including North Korea—that are already developing longer-range systems. We also have an obligation to our allies, some of whom have signed on as partners in this effort. The department is proceeding with negotiations with Poland and the Czech Republic on establishing a missile defense capability in Europe while we work with our other allies, including the United Kingdom, on upgrading early warning radar systems. We are willing to partner with others in developing this defensive capability, including Russia. The missile defense program funded by this request will continue to test our capability against more complex and realistic scenarios. I urge the committee to approve the full $9.9 billion requested for the missile defense and Patriot missile programs.

## Space Capabilities

The recent test of an anti-satellite weapon by China underscored the need to continue to develop capabilities in space. The policy of the U.S. Government in this area remains consistent with the long-standing principles that were established during the Eisenhower administration, such as the right of free passage and the use of space for peaceful purposes. Space programs are essential to the U.S. military's communications, surveillance, and reconnaissance capabilities. The base budget requests about $6.0 billion to continue the development and fielding of systems that will maintain U.S. supremacy while ensuring unfettered, reliable, and secure access to space.

## Recapitalization

A major challenge facing our military is that several key capabilities are aging and long overdue for being replaced. The prime example is the Air Force KC-135 tanker fleet. With planes that average 45 years of age, the fleet is becoming more expensive to maintain and less reliable to operate. The Air Force has resumed a transparent and competitive replacement program to recapitalize this fleet with the KC-X aircraft. The KC-X will be able to carry cargo and passengers and will be equipped with defensive systems. It is the U.S. Transportation Command's and the Air Force's top acquisition and recapitalization priority.

## Train and Equip Authorities

Recent operations have shown the critical importance of building the capacity and capability of partners and allies to better secure and govern their own countries. In recent years we have struggled to overcome the patchwork of authorities and regulations that were put in place during a very different era—the Cold War—to confront a notably different set of threats and challenges.

The administration has, with congressional support, taken some innovative steps to overcome these impediments. A significant breakthrough was the Section 1206 authority, which fills a critical gap between traditional security assistance and direct U.S. military action. It allows the Defense and State Departments to build partner nations' security capacity in months, rather than years. The program focuses on capacity-building in places where we are not at war, but face emerging threats or opportunities. DoD and State cooperation in executing this program has been excellent and serves as a model for developing other whole-of-government approaches to complex security problems.

Section 1206 projects approved last year are already helping partners reduce threats to global resource flows, narrow terrorists' freedom of action, and increase stability in sensitive regions. The Chairman of the Joint Chiefs and the combatant commanders regard this program as the most important authority the military has to fight the War on Terror beyond Iraq and Afghanistan, because it allows us to help others get ahead of threats, exploit opportunities, and reduce stress on our active duty, reserve and National Guard servicemen and women.

For FY 2007, combatant commanders and country teams have identified nearly $800 million in projects globally. We sought $300 million in the Supplemental and are seeking dedicated funding of $500 million in the FY 2008 base budget to provide the combatant commanders with the resources to implement this authority.

Building the capacity and capability of partners and allies to better secure and govern their own countries is a central task to counter terrorism. Dedicated funding will help us accomplish this task without disrupting other vital DoD programs. It is much more effective for partner countries, rather than US forces, to defeat terrorists operating within their borders. We strongly urge your support for this critical program.

### Quality of Life—Sustaining the All-Volunteer Force

Our nation is fortunate that so many talented and patriotic young people have stepped forward to serve, and that so many of them have chosen to continue to serve. So far, all active branches of the U.S. military exceeded their recruiting goals, with particularly strong showings by the Army and Marine Corps. The FY 2008 request includes $4.3 billion for recruiting and retention to ensure that the military continues to attract and retain the people we need to grow the ground forces and defend the interests of the United States.

We will continue to support the all-volunteer force and their families through a variety of programs and initiatives. The budget includes:
- $38.7 billion for health care for both active and retired service members;
- $15 billion for Basic Allowance for Housing to ensure that, on average, troops are not forced to incur out-of-pocket costs to pay for housing;
- $2.9 billion to improve barracks and family housing and privatize an additional 2,870 new family units; and
- $2.1 billion for a 3 percent pay increase for military members.

In addition, recently announced changes in the way the military uses and employs the Reserves and National Guard should allow for a less frequent and more predictable mobilization schedule for our citizen soldiers.

Combined with other initiatives to better organize, manage, and take care of the force, these changes should mean that in the future our troops should be deployed or mobilized less often, for shorter periods of time, and with more predictability and a better quality of life for themselves and their families.

### Global War on Terror Request

The President's FY 2008 Global War on Terror Request for $141.7 billion complies with Congress's direction to include the costs of ongoing operations in

Iraq and Afghanistan in the annual Defense Department budget. Given the uncertainty of projecting the cost of operations so far in the future, the funds sought for the FY 2008 GWOT Request are generally based on a straight-line projection of current costs for Iraq and Afghanistan. This request includes $70.6 billion to provide the incremental pay, supplies, transportation, maintenance and logistical support to conduct military operations.

### Reconstitution

The FY 2008 GWOT Request includes $37.6 billion to reconstitute our nation's armed forces—in particular, to refit the ground forces, the Army and Marine Corps, who have borne the brunt of combat in both human and material terms. These funds will go to repair or replace equipment that has been destroyed, damaged, or stressed in the current conflict. In many cases, reconstitution funds will provide upgraded and modernized equipment to replace older versions. The $13.6 billion in reset funds in the FY 2008 GWOT Request for the U.S. Army will go a long way towards replacing items, one for one, that were worn out or lost during operations to ensure force readiness remains high.

### Force Protection

This FY 2008 GWOT Request includes $15.2 billion for investments in new technologies to better protect our troops from an agile and adaptive enemy. Programs being funded would include a new generation of body armor, vehicles that can better withstand explosions from Improvised Explosive Devices (IEDs), and electronic devices that interrupt the enemy's ability to attack U.S. forces. Within this force-protection category, the FY 2008 GWOT Request includes $4 billion to counter and defeat the threat posed by IEDs.

### Afghan/Iraqi Security Forces

The FY 2008 GWOT Request includes $4.7 billion to stand up capable military and police forces in Afghanistan and Iraq. The bulk of these funds are going to train and equip Afghan National Security Forces (ANSF) to assume the lead in operations throughout Afghanistan. As of February, over 90,000 had been trained and equipped, an increase of more than 33,000 from the previous year.

In Iraq, approximately 334,000 soldiers and police have been trained and equipped, and are in charge of more than 60 percent of Iraqi territory and more than 65 percent of that country's population. They have assumed full security responsibility for four out of Iraq's 18 provinces and are scheduled to take over more territory over the course of the year. These Iraqi troops, though far from perfect, have shown that they can perform with distinction when properly led and supported.

## Non-Military Assistance

Success in the kinds of conflicts our military finds itself in today—in Iraq, or elsewhere—cannot be achieved by military means alone. The President's strategy for Iraq hinges on key programs and additional resources to improve local governance, delivery of public services, and quality of life—to get angry young men off the street and into jobs where they will be less susceptible to the appeals of insurgents or militia groups.

Commanders Emergency Response Program (CERP) funds are a relatively small piece of the war-related budgets—$977 million in the FY 2008 GWOT Request. But because they can be dispensed quickly and applied directly to local needs, they have had a tremendous impact—far beyond the dollar value—on the ability of our troops to succeed in Iraq and Afghanistan. By building trust and confidence in Coalition forces, these CERP projects increase the flow of intelligence to commanders in the field and help turn local Iraqis and Afghans against insurgents and terrorists.

## Conclusion

With the assistance and the counsel of Congress, I believe we have the opportunity to do right by our troops and the sacrifices that they and their families have made these past few years. That means we must make the difficult choices and commit the necessary resources to not only prevail in the current conflicts in which they are engaged, but to be prepared to take on the threats that they, their children, and our nation may face in the future.

## Team America Rocketry Challenge

*The Plains, Virginia, Saturday, May 19, 2007*

THANK you John. And it is a pleasure to be here. I appreciate the kind introduction. And thanks to the Aerospace Industries Association and the National Association of Rocketry for inviting me to this great event.

Texas A&M is the sixth largest university in the country, with 46,000 students. Ten thousand of them are in engineering. I have a feeling that your secret is out, and that you may find your ranks swollen next year by some recruiters from various universities.

Well I just came from seeing some very impressive displays. And congratulations to everyone for the hard work that you put into preparing for today. The awards for the challenge, as you know, are going to be given out in a few minutes. It's perfectly obvious, I think, why they make you sit through the speeches before the awards are given out.

There is a large group of parents, coaches, and teachers here, and I especially wanted to recognize you. You are supporting your children and young adults in an activity that's not only educational but obviously a lot of fun.

As Secretary of Defense, I'm in charge of some of the most high-tech hardware anywhere in the world. One of the most advanced projects is a system to shoot down missiles that might be fired at our country. Basically, it's like trying to hit a bullet with another bullet. It's not easy. And if any of you had a breakthrough on propulsion or ballistics during this competition, I hope you'll let me know.

The work of the Department of Defense protects our nation but it also benefits the public in ways you may not realize. Some of you may have used GPS to get here. And then there's the Internet. Not everyone knows that these technologies had their origins in Department of Defense research.

You've heard that invention is one percent inspiration and 99 percent perspiration. That recipe leaves out one crucial ingredient—the eggs. Even when they did break over the last couple of days, a valuable lesson was reinforced: that an experiment that doesn't go perfectly is not failure. Each attempt is one step along the way to making something work better.

The task you were given has a wonderful historical echo to it. Your goal was to propel a raw egg to a height of 850 feet and, within 45 seconds, return it safely to earth. I remember when President Kennedy challenged America to "commit itself to achieving the goal, before this decade is out, of landing a man on the moon and returning him safely to earth."

I'd like to talk about someone else from that era for just a couple of minutes, who no doubt broke a few eggs in his time. His name was Homer Hickam. You've probably heard of him, or seen the movie "October Sky."

Homer grew up during the 1950s in a poor mining town in West Virginia. People my age recall when the Soviet Union sent the first satellite, called Sputnik, into space in 1957. While Sputnik made the parents of America worry that the United States was falling behind the Russians in science, some of their children, like Homer Hickam, felt inspired—inspired to send something into space themselves.

To do this, Homer and his buddies had to start from scratch. They went searching through the kitchen drawer, the company store, and the local library, for materials and know-how. They used cardboard, model glue, saltpeter, scrap metal, and the application of Newton's third law of action and reaction.

The first rocket Homer made blew up and destroyed his mother's fence. It took the team several tries to get a rocket to rise six feet.

At first, the "rocket boys" were treated as the village nuisance. But it didn't stop them. They built their own rocket range. They called it Cape Coalwood, in imitation of Cape Canaveral. Eventually, their parents and the entire community rallied behind them.

When they competed at science fairs, they were up against hotshots from bigger schools and bigger towns. But Homer and his friends won prizes anyway. And their efforts took them places they never imagined they would go.

Each of them went to college, in a community where that was rare. Four became engineers. Homer went on to work for NASA, where he helped design many of America's spacecraft.

I tell you this story because it's about becoming part of something larger than yourself. It's about friendship and the excitement of accomplishing something difficult with others who are as passionate about the same things as you are.

Your teamwork on your rockets has done this for you as well. Today's contest and others like it are fun, and they're great learning experiences. They can also be the beginning of a journey that will lead you as far as your aspirations can take you. Science opens up our world to us—and it is a breath-taking place. There has never been a better time to learn about physics, the mechanics of flight, and space exploration.

You've taken your first steps; now keep going. As Virgil said, "So shall you scale the stars!"

Thank you.

## College of William and Mary Commencement

*Williamsburg, Virginia, Sunday, May 20, 2007*

THANK you, President Nichol. Members of the faculty, parents, distinguished guests. Justice O'Connor—Chancellor—a pleasure to see you. Justice O'Connor administered my oath of office as Director of Central Intelligence in 1991 and, more recently, as President Nichol has mentioned, we served on the Baker-Hamilton Commission last year—although my tenure on the group was rather abruptly interrupted.

Speaking of which, in terms of my timing in taking on the responsibilities of the Secretary of Defense, it reminds me of a story told long ago by Senator Richard Russell of Georgia, who spoke of having seen a bull that charged a locomotive. He said, "You know that was the bravest bull I ever saw, but I can't say much for his judgment."

Dr. Kelso and Secretary Coleman, your recognition here today is well-deserved.

To the members of the Class of 2007: Congratulations. I am truly honored—and flattered—to be your graduation speaker.

I presided over 39 commencement ceremonies as president of Texas A&M, yet, today is the first commencement speech I have ever given. I thank all of you for the extraordinary privilege of letting it be at my alma mater.

To the parents: you must be welling up with pride at the achievements of your children. Having put two children through college, I know there are many sighs of relief as well, and you are probably already planning how to spend your newly re-acquired disposable income. Forget it. Trust me on this. If you think you've written your last check to your son or daughter, dream on. The National Bank of Mom and Dad is still open for business.

I guess I am supposed to give you some advice on how to succeed. I could quote the billionaire J. Paul Getty, who offered advice on how to get rich. He said, "Rise early, work late, strike oil." Or, Alfred Hitchcock, who said, "There's nothing to winning really. That is if you happen to be blessed with a keen eye, an agile mind, and no scruples whatsoever."

Well, instead of those messages, my only words of advice for success today comes from two great women. First, opera star Beverly Sills, who said,

"There are no short cuts to anyplace worth going." And second, from Katharine Hepburn, who wrote: "Life is to be lived. If you have to support yourself, you had bloody well find some way that is going to be interesting. And you don't do that by sitting around wondering about yourself."

In all those 39 commencements at Texas A&M, I learned the importance of brevity for a speaker. George Bernard Shaw once told a speaker he had 15 minutes. The speaker asked, "How can I possibly tell them all I know in 15 minutes? Shaw replied, "I advise you to speak very slowly." I will speak quickly, because, to paraphrase President Lincoln, I have no doubt you will little note nor long remember what is said here.

I arrived at William & Mary in 1961 at age 17, intending to become a medical doctor. My first year was pure pre-med: biology, chemistry, calculus and so on. I soon switched from pre-med to history. I used to say "Godonly knows how many lives have been saved by my becoming Director of CIA instead of a doctor."

When reflecting on my experience here I feel gratitude for many things:

- To William & Mary for being a top-tier school that someone like me could actually afford to attend—even as an out-of-state student. By the way, hold on to your hats, parents: Out of state tuition then was $361 a semester.
- Gratitude for the personal care and attention from a superb faculty and staff—a manifestation of this university's commitment to undergraduate education that continues to this day;
- Gratitude to those in the greater Williamsburg community, who opened their hearts and their homes to a 17-year-old far from his own home; and
- Gratitude for one more thing. During my Freshman year I got a 'D' in calculus. When my father called from Kansas to ask how such a thing was possible, I had to admit, "Dad, the 'D' was a gift." So, I'm grateful to that math professor too.

What William & Mary gave me, above all else, was a calling to serve—a sense of duty to community and country that this college has sought to instill in each generation of students for more than 300 years. It is a calling rooted in the history and traditions of this institution.

Many a night, late, I'd walk down Duke of Gloucester Street from the Wren Building to the Capitol. On those walks, in the dark, I felt the spirit of the patriots who created a free and independent country, who helped birth it right here in Williamsburg. It was on those walks that I made my commitment to public service.

I also was encouraged to make that commitment by the then-president of the United States, John F. Kennedy, who said to young Americans in the early 1960s, "Ask not what your country can do for you, but ask what you can do for your country."

We are celebrating the 400th anniversary of the founding of Jamestown. Looking back, it's hard to imagine this country could have gotten off to a more challenging start. It began as a business venture of a group of London merchants with a royal patent. The journalist Richard Brookhiser recently compared it to Congress today granting Wal-Mart and GE a charter to colonize Mars.

Brookhiser wrote,"Its leaders were always fighting. Leaders who were incompetent or unpopular—sometimes the most competent were the least popular—were deposed on the spot," He continues, "The typical 17th Century account of Jamestown argues that everything would have gone well if everyone besides the author had not done wrong." Sounds like today's memoirs by former government officials.

Jamestown saw the New World's first representative assembly—the institutional expression of the concept that people should have a say in how they were governed, and having that say brought with it certain obligations: a duty to participate, a duty to contribute, a duty to serve the greater good.

It is these four-hundred-year-old obligations that I want to address for the next few minutes. When talking about American democracy, we hear a great deal about freedoms, and rights, and, more recently, about the entitlements of citizenship. We hear a good deal less about the duties and responsibilities of being an American.

Young Americans are as decent, generous, and compassionate as we've ever seen in this country—an impression reinforced by my four and a half years of experience as President of Texas A&M, by the response of college students across America—and especially here at William & Mary—to the tragedy at Virginia Tech, and even more powerfully reinforced by almost six months as Secretary of Defense.

That is what makes it puzzling that so many young people who are public-minded when it comes to their campus and community tend to be uninterested in— if not distrustful of—our political processes. Nor is there much enthusiasm for participating in government, either as a candidate or for a career.

While volunteering for a good cause is important, it is not enough. This country will only survive and progress as a democracy if its citizens—young and old alike—take an active role in its political life as well.

Seventy percent of eligible voters in this country cast a ballot in the election of 1964. The voting age was then 21. During the year I graduated, 1965, the first major American combat units arrived in Vietnam, and with them, many 18-, 19-, and 20-year-olds. In recognition of that disparity, years later the voting age would be lowered to 18 by constitutional amendment.

Sad to say, that precious franchise, purchased and preserved by the blood of hundreds of thousands of Americans your age and younger from 1776 to today, has not been adequately appreciated or exercised by your generation.

In 2004, with our nation embroiled in two difficult and controversial wars, the voting percentage was only 42 percent for those aged 18 to 24.

Ed Muskie, former senator and Secretary of State, once said that "you have the God given right to kick the government around." And it starts with voting, and becoming involved in campaigns. If you think that too many politicians are feckless and corrupt, then go out and help elect different ones. Or go out and run yourself. But you must participate, or else the decisions that affect your life and the future of our country will be made for you—and without you.

So vote. And volunteer. But also consider doing something else: dedicating at least part of your life in service to our country.

I entered public life more than 40 years ago, and no one is more familiar with the hassles, frustrations and sacrifices of public service than I am. Government is, by design of the Founding Fathers, slow, unwieldy and almost comically inefficient. Will Rogers used to say: "I don't make jokes. I just watch the government and report the facts."

These frustrations are inherent in a system of checks and balances, of divisions and limitations of power. Our Founding Fathers did not have efficiency as their primary goal. They designed a system intended to sustain and protect liberty for the ages. Getting things done in government is not easy, but it's not supposed to be.

I last spoke at William & Mary on Charter Day in 1998. Since then our country has gone through September 11 with subsequent wars in Afghanistan and Iraq. We learned once again that the fundamental nature of man has not changed, that evil people and forces will always be with us, and must be dealt with through courage and strength.

Serving the nation has taken on a whole new meaning and required a whole new level of risk and sacrifice—with hundreds of thousands of young Americans in uniform who have stepped forward to put their lives on the line for their country. These past few months I've met many of those men and women—in places like Fallujah and Tallil in Iraq and Bagram and Forward Operating Base Tillman in Afghanistan—and at Walter Reed as well. Seeing what they do every day, and the spirit and good humor with which they do it, is an inspiration. The dangers they face, and the dangers our country faces, make it all the more important that this kind of service be honored, supported, and encouraged.

The ranks of these patriots include the graduates of William & Mary's ROTC program, and the cadets in this Class of 2007, who I'd like to address directly. You could have chosen a different path—something easier, or safer, or better compensated—but you chose to serve. You have my deepest admiration and respect—as Secretary of Defense, but mostly as a fellow American.

You are part of a tradition of voluntary military service dating back to George Washington's Continental Army. That tradition today includes General David McKiernan, William & Mary Class of 1972, who led the initial ground force in Iraq and now commands all Army troops in Europe. It also is a tradition not without profound loss and heartache.

Some of you may know the story of Ryan McGlothlin, William & Mary Class of 2001: a high school valedictorian, Phi Beta Kappa here, and Ph.D. candidate at Stanford. After being turned down by the Army for medical reasons, he persisted and joined the Marines and was deployed to Iraq in 2005. He was killed leading a platoon of riflemen near the Syrian border.

Ryan's story attracted media attention because of his academic credentials and family connections. That someone like him would consider the military surprised some people. When Ryan first told his parents about joining the Marines, they asked if there was some other way to contribute. He replied that the privileged of this country bore an equal responsibility to rise to its defense.

It is precisely during these trying times that America needs its best and brightest young people, from all walks of life, to step forward and commit to public service. Because while the obligations of citizenship in any democracy are considerable, they are even more profound, and more demanding, as citizens of a nation with America's global challenges and responsibilities—and America's values and aspirations.

During the war of the American Revolution, Abigail Adams wrote the following to her son, John Quincy Adams: "These are times in which a genius would wish to live. It is not in the still calm of life, or the repose of a pacific station that great characters are formed... Great necessities call out great virtues."

You graduate in a time of "great necessities." Therein lies your challenge and your opportunity.

A final thought. As a nation, we have, over more than two centuries, made our share of mistakes. From time to time, we have strayed from our values; and, on occasion, we have become arrogant in our dealings with others. But we have always corrected our course. And that is why today, as throughout our history, this country remains the world's most powerful force for good—the ultimate protector of what Vaclav Havel once called "civilization's thin veneer." A nation Abraham Lincoln described as mankind's "last, best hope."

If, in the 21stcentury, America is to be a force for good in the world—for freedom, the rule of law, and the inherent value of each and every person; if

America is to continue to be a beacon for all who are oppressed; if America is to exercise global leadership consistent with our better angels, then the most able and idealistic of your generation must step forward and accept the burden and the duty of public service. I promise you that you will also find joy and satisfaction and fulfillment.

I earlier quoted a letter from Abigail Adams to her son, John Quincy. I will close with a quote from a letter John Adams sent to one of their other sons, Thomas Boylston Adams. And he wrote: "Public business, my son, must always be done by somebody. It will be done by somebody or another. If wise men decline it, others will not; if honest men refuse it, others will not."

Will the wise and the honest among you come help us serve the American people?

Congratulations and Godspeed.

# United States Naval Academy Commencement

*Annapolis, Maryland, Friday, May 25, 2007*

THANK you, Secretary Winter.

Admiral Mullen, General Magnus, distinguished guests. It is a special honor to join you today for this long-anticipated and well-deserved celebration. It is an honor I thought I would never have, and perhaps for more than a few of you, a celebration you thought would never come.

While I presided over 39 commencement ceremonies as President of Texas A&M University, this is only the second one at which I have spoken. One lesson I learned from those 39 commencements was to keep it short. Because, to paraphrase President Lincoln, you will little note nor long remember what is said here.

I want to welcome and thank family members who are here. Your support and encouragement have made this day possible for these young men and women. More importantly, you have nourished their spirits and molded their character. You have instilled in them love of country and a willingness to serve. And now you entrust to the nation your most treasured possession. Words cannot express our gratitude or admiration for your accomplishment—your success manifested in the quality of these young people. They will continue to need your support throughout their careers.

Thanks also to the sponsor families of these midshipmen. Over the past four years, you have opened your homes to these young men and women, providing a good meal or a respite from Academy life. Or a shoulder to lean on. Your guidance and your caring helped make today possible for your mids.

This has been quite a couple of weeks, with visits from the Blue Angels, your last parade on Worden Field, and finally escaping Bancroft Hall. I am sure you watched plebes climbing Herndon, and felt obligated to offer casual observations on how your class did it better. The memories of this past week—the past four years for that matter—probably seem like a whirlwind.

Of course, your class is no stranger to strong winds. Three months after you arrived here in 2003, Hurricane Isabel paid Annapolis a visit. There is a saying about how midshipmen spend their "four years together by the bay." I guess you had a short time to spend together in the bay.

Speaking of adventures—or should I say misadventures—I want to exercise my authority as Secretary of Defense to grant amnesty to all midshipmen whose antics led to minor conduct offenses. As always, Admiral Rempt has the final say on what constitutes "minor."

I understand we have quite a few budding aviators and flight officers in this group. Right now, aircraft from the carriers USS Nimitz and USS Stennis are in the skies over the Middle East, watching over their comrades on the ground. It is an awesome responsibility—one I hope you embrace each time you catapult off the end of a carrier.

For you surface warriors, as has been the case since the earliest days of our republic, you will carry our nation's sovereignty—and protect our nation's power—to the far reaches of the globe. The mere sight of you will instill caution into those who would threaten peace, or, in some cases, bring hope to those suffering in the wake of natural disasters—missions that display our nation's resolve and communicate our nation's values.

To our submariners: for much of my life, submarines existed mostly to confront the Soviet Navy, and to be America's strategic response in the event of a nuclear war—services our nation hoped we would never need. The sub fleet you will soon join is actively engaged in a range of missions—from reconnaissance to special operations. And yes, extremists on remote desert mountaintops are quite surprised to find themselves coming under fire by American submarine crews.

Last, but certainly not least, our newest "Devil Dogs." Last month, General Pace and I had the opportunity to meet with Marines in Fallujah. As you know, Fallujah was once a terrorist stronghold that became the site of some of the harshest close combat our military has seen in decades. Like in so many places before in the Corps' proud history, Marines today are giving Fallujah residents a chance for a brighter future. With every action you take as Marines, with every person you lead, you will build on the legacy of faithful service that has defined the United States Marine Corps for over two centuries.

I have four points I want to make this morning.

First, I want to thank you for choosing to serve your country and your fellow citizens. In everything you did here—from studying for exams to training sessions with your upperclassmen—you have grown together as a team. But there has also been something bigger uniting you: your willingness to take on a difficult and dangerous path in the service of others. Your class motto is fitting: "Libertas per Sacrificium," Liberty through Sacrifice.

Most of you were juniors in high school when terrorists attacked America in September 2001, and it became clear we were a nation at war. With your credentials, you could have attended another prestigious university, and subsequently pursued a private life, with all of its material rewards, your freedom and safety assured by other young men and women who volunteered to serve in the American military.

You, however, are special, because you are among those who have chosen to serve—to defend the dreams of others. And that sets you apart.

In the latest of nights—while studying for naval leadership finals—you may have wondered why you needed to know the differences between Chester Nimitz and Chesty Puller. But in studying the great Navy and Marine Corps leaders—and the people and battles memorialized throughout this stadium—you learned the one thing that united all of the men and women who came before you: they served because they loved their country more than themselves.

While many people may witness history, those who step forward to serve in a time of crisis never have to question their place in history. And, as today you join the long line of patriots in a noble calling, by your service you will have your chance to make history.

During the War of the American Revolution, Abigail Adams wrote the following to her son, John Quincy: "These are times in which a genius would wish to live. It is not in the still calm of life or [in] the repose of a pacific station that great characters are formed. …great necessities call out great virtues."

You begin your service in a time of "great necessities." Therein lies your challenge and your opportunity. And for your willingness to serve, the entire nation is grateful.

My second point. Today you will take an oath to protect and defend the Constitution of the United States. I have taken that oath seven times in the last forty years—the first when I enlisted in 1966 and the last when I became Secretary of Defense.

Today, I want to encourage you always to remember the importance of two pillars of our freedom under the Constitution—the Congress and the press. Both surely try our patience from time to time, but they are the surest guarantees of the liberty of the American people.

The Congress is a co-equal branch of government that under the Constitution raises armies and provides for navies. Members of both parties now serving in Congress have long been strong supporters of the Department of Defense, and of our men and women in uniform.

As officers, you will have a responsibility to communicate to those below you that the American military must be non-political and recognize the obligation we owe the Congress to be honest and true in our reporting to them. Especially when it involves admitting mistakes or problems.

The same is true with the press, in my view a critically important guarantor of our freedom. When it identifies a problem, as at Walter Reed, the response of senior leaders should be to find out if the allegations are true—as they were at Walter Reed—and if so, say so, and then act to remedy the problem. If untrue, then be able to document that fact. The press is not the enemy, and to treat it as such is self-defeating.

As the Founding Fathers wisely understood, the Congress and a free press, as with a non-political military, assure a free country. A point underscored by a French observer writing about George Washington in 1782. He wrote: "This is the seventh year that he has commanded the army and that he has obeyed the Congress; more need not be said."

A third point. Don't ever lose your sense of humor. In times of unbearable stress and crisis, finding a way to smile or laugh will make you a better officer and a better decision-maker. So, once in a while, put your trash can on your desk and label it "in-box;" when the money comes out of the ATM, yell "I won, I won;" at lunch time, sit in your parked car with sunglasses on and point a hair dryer at passing cars and see if they slow down.

I'm kidding, but you get the point. Above all, be willing to laugh at yourself. The bottom line: humor will keep you sane and balanced in the most difficult circumstances.

Fourth, and finally, I want to say a few words about leadership, a subject you have studied at length here at the Academy and learned through experience as midshipmen. I probably can't tell you anything new about leadership, but perhaps I can offer a different perspective.

You see, I believe real leadership is a rare commodity. Believe me, I know. Over the course of my career, I have been privileged to work for seven presidents. I knew six personally and worked for four in the White House. I have worked with 10 secretaries of state, 8 secretaries of defense, 11 national security advisors, 9 chairmen of the Joint Chiefs of Staff, and more admirals, generals, and ambassadors than I can count. I witnessed in person and in action world figures such as Margaret Thatcher, Mikhail Gorbachev, Boris Yeltsin, Francois Mitterand, Helmut Kohl, Lech Walesa, Anwar Sadat, Yitzhak Rabin, and many others. I suppose I am sort of a global Forrest Gump.

To be a leader at any level, from president to junior officer, I believe the following qualities are required:

Vision. If you would be a real leader, you must see beyond the day-to-day tasks and challenges. You must look beyond tomorrow and discern a world of possibilities and potential. You must see what others do not or cannot and then be prepared to act on your vision.

Another quality is integrity. Without this, real leadership is not possible. Nowadays, it seems like integrity—or honor or character—is kind of quaint, a curious, old-fashioned notion.

But there are many, many people for whom personal integrity and honor are as important as life itself. I have encountered many such people during my forty years in public service—and you will encounter many during your careers.

For a real leader, personal virtues—self-reliance, self-control, honor, truthfulness, morality—are absolute. These are the building blocks of character, of integrity—and only on that foundation can true leadership be built.

An additional quality necessary for leadership is deep conviction. True leadership is a fire in the mind that transforms all who feel its warmth, that transfixes all who see its shining light in the eyes of a man or woman. It is a strength of purpose and belief in a cause that reaches out to others, touches their hearts, and makes them eager to follow.

Self-confidence is still another quality of leadership. Not the chest-thumping, strutting egotism we see and read about all the time. Rather, it is the quiet self-assurance that allows a leader to give others both real responsibility and real credit for success. The ability to stand in the shadow and let others receive attention and accolades. A leader is able to make decisions but then delegate and trust others to make things happen.

This doesn't mean turning your back after making a decision and hoping for the best. It does mean trusting people at the same time you have a regular reporting mechanism, and are holding them accountable. The bottom line: a self-confident leader doesn't cast such a large shadow that no one else can grow.

A further quality of leadership is courage: the courage to chart a new course; the courage to do what is right and not just what is popular; the courage to stand alone; the courage to act; the courage as a military officer to "speak truth to power."

In most academic curricula today, and in most business, government, and military training programs, there is great emphasis on team-building, on working together, on building consensus, on group dynamics. You have heard a lot about that. But, for everyone who would become a leader, the time comes when he or she must stand alone and say, "This is wrong" or "I disagree with all of you and, because I have the responsibility, this is what we will do." Don't kid yourself—that takes courage.

Vision, integrity, deep conviction, and self-confidence are not enough to make a leader; a leader must have the courage to act, often against the will of the crowd. As President Ford once said, "... the greatest defeat of all would be to live without courage, for that would hardly be living at all."

A final quality of real leadership, I believe, is simply common decency: treating those around you—and, above all, your subordinates—with fairness and respect. An acid test of leadership is how you treat those you outrank.

A real leader, in my experience, from the lowest rung of the ladder to the top, treats every person with respect and dignity.

Use your authority over others for constructive purposes, to help them—to watch out and care for them, to help them improve their skills and to advance, to ease their hardships whenever possible. All of this can be done without compromising discipline or mission or authority.

Common decency builds respect and, in a democratic society, respect is what prompts people to give their all for a leader, even at personal sacrifice.

In a novel about ancient Greece, the warrior Alcibiades is asked how to lead free men, and he responds: "By being better and thus commanding their emulation." Alcibiades goes on, "A commander's role is to model ... excellence before his men. One need not thrash them to greatness, only hold it out before them. They will be compelled by their own nature to emulate it." He concludes: "How [to] lead free men? Only by this means: the summoning of each to his nobility."

Today is a joyous day in a difficult time. So thank your parents and your sponsor families for all the love and care they have given you. And celebrate. For today, you take on the awesome responsibility of protecting and defending the Constitution of the United States and the American people. Today, we ask you to make the extraordinary the expected. Today, we ask you to lead free men and women by summoning each to his or her nobility.

Congratulations, God bless, and Godspeed.

## Air Force Academy Commencement

*Colorado Springs, Colorado, Wednesday, May 30, 2007*

THANK you, Secretary Wynne.
General Moseley, Senator Bunning. Distinguished guests, members of the public, leaders of the Air Force—past, present, and future.
One of the great advantages of being Secretary of Defense is that I have many opportunities to interact with our military's top leaders—and I even have quite a few opportunities to pay tribute to them publicly when they retire or move on to new commands.
I don't, however, have nearly as many chances to pay tribute to our nation's youngest leaders—to thank them for their service. So I am grateful for the chance to thank the Class of 2007 for their choice.
And I am particularly honored to be able to say to the Class of 2007: Congratulations on this achievement! You've certainly earned it.
I presided over 39 commencement ceremonies as president of Texas A&M University. One thing I learned from 39 commencement addresses: keep it short. Because to paraphrase President Abraham Lincoln, "you will little note nor long remember" what is said here today.
George Bernard Shaw once told a speaker he had 15 minutes to speak. The speaker replied, "15 minutes? How can I tell them all I know in 15 minutes?" Shaw responded, "I advise you to speak very slowly." I'll try to speak quickly.
On the way out here, my Air Force pilots had a chance to give me a few minor suggestions for my speech.
They said that I should definitely mention Billy Mitchell and Curtis LeMay and Hap Arnold.
And Eddie Rickenbacker and Dick Bong and Steve Ritchie.
And the Doolittle Raiders and the Flying Tigers and the Tuskegee Airmen.
And even Old Chicago's and Phantom Canyon and Hap's Place. I told them I wasn't sure if I could work them all in, but I'd do my best. They responded by telling me they weren't sure if they could avoid heavy turbulence on the flight home, but they'd certainly do their best.

I was commissioned as an Air Force second lieutenant in 1966. By that time, the importance of airpower was an accepted fact. But people forget that it wasn't always so.

In the early 1900s, right after the dawn of flight, other services had no great love of airplanes. The cavalry in particular was opposed to their use, because they were afraid the planes would scare the horses.

The intellectuals weren't much better. In 1908, one of the nation's leading astronomers wrote in the trade magazine Aeronautics that—and I quote—"another popular fallacy is to suppose that flying machines could be used to drop dynamite on an enemy in [a] time of war."

Even the guys who invented the airplane had their share of trouble. On a test flight for an Army contract, Orville Wright took to the skies with Lieutenant Thomas Selfridge, who served on the Army's contract board.

After a few minutes in the air, disaster struck and the plane crashed to the ground. Lieutenant Selfridge's last words were, "Take this damn thing off my back."

I should note that the Wright brothers eventually got the Army contract.

Since then and throughout the 60-year history of the Air Force, Americans have stood in awe as airmen pushed the limits of technology and courage. Airmen have crashed through the sound barrier many times over, and extended the range, scope, and nature of air missions beyond what anyone could have imagined—to the point of running 7,000-mile B-2 bombing sorties in Iraq from Whiteman Air Force Base, where I was assigned to a Minuteman missile wing more than 40 years ago.

It is upon this great tradition of technological innovation that the Air Force was formed, and it is this great legacy of personal courage that lives within each and every graduate of this institution.

Four years ago you joined the Air Force Academy as lowly fourth classmen. Today you leave as officers in the United States Air Force—the sword and shield of the United States, its sentries and its avengers.

So it's not an easy path. You are one of the first classes to begin the arduous process of application to the Academy since September 11th. You knew the dangers of the world you were entering, but you still chose to step forward. You still chose to embark on the journey that brings us here today.

Along the way, you learned a tremendous amount about the Air Force—its history, its traditions, its great personalities, its evolution, its future.

But, just as important, you learned a tremendous amount about yourselves. About your commitment. About your endurance. About what it takes to be a leader.

There were also many successful ventures outside the classroom—from setting a world record for a free-fall formation, to having the best basketball season in the Academy's history, to learning about the dangers of dancing alone in your dorm room if your roommate happens to own a video camera. And yes, I've seen the video. Don't give up your day job.

Even embarrassment is part of growing up.

And forcing all of you to grow up—to live up to the highest standards personally and professionally—is a large part of what the Academy has done over the past four years. This institution strives to create not just a better officer, but a better individual—even if at times it seemed unclear exactly how SAMIs and rifle runs would do that.

But we are not here today just to look back to the past and recognize all that you have achieved. We're also here to look ahead to the future—to recognize all that you will achieve in the coming years.

It is by no means an easy future. We are engaged in two wars on the other side of the world—and we are engaged in a global ideological struggle against some of the most barbaric enemies we have ever faced. There are also many threats on the horizon, both traditional and non-traditional, and as always there are the threats that still lie beyond the horizon, threats we cannot yet even perceive.

The world you are entering is much more complicated than it was when I was an Air Force officer during the Cold War. You will not always know who your enemies are. You will not always be able to understand their motivations. And you will not always be able to rely exclusively on technology to win battles or wars.

The challenges you face will test both your spirit and your resolve. At the Academy you have undoubtedly heard much about what it takes to be a leader. Well, the time for words has now passed. From this day forward, you will have to demonstrate that you can live up to the standards you were taught. That you can perform in a military that is unique in the world in terms of how heavily it relies on the judgment and integrity of its junior officers.

I can tell you that it will rarely, if ever, be easy. Far too often today we see the results of a failure of leadership at too many levels—whether in the home, in schools, in business, in government, and yes, even in the armed forces. It certainly does not have to do with the natural capabilities of our leaders. They are for the most part smart, educated, driven. They did not rise to positions of leadership by accident—but by demonstrating a capacity, and a willingness, to lead.

But the ability to lead carries with it great responsibilities—for it is just as easy, if not easier, to lead people down the wrong path. It is easy to try to cover your tracks if you make a mistake.

It is easy to give your superiors good news even when you know it is not warranted. It is all too easy to sacrifice the long-term interests of the service and the nation for short-term personal gains.

All these things are easy, but they're wrong. Moral quandaries of the sort you will face are made more difficult by the realities of the world today.

We live in an age where friends and enemies alike will seek out and focus on any and all mistakes made under great stress; where the irregular battlefield will present life-and-death decisions, often with no good choices. Where the slightest error in judgment—or even the perception of an error—can be magnified many times over on the Internet and on TV and circulated around the globe in seconds.

You will face enemies who possess no conscience and no remorse—who will lie about and distort your actions, and who will purposefully blur the line between civilians and combatants.

Your actions will also be under scrutiny by those who support you—by the Congress, the press, and by everyday citizens. And make no mistake about it—your supporters at home will be watching—and setting their expectations high.

There is only one way to conduct yourself in this world—only one way to remain always above reproach. For a real leader, the elements of personal virtue—self-reliance, self-control, honor, truthfulness, morality—are absolute. They are absolute even when doing what is right may bring embarrassment or bad publicity to your unit or the service or to you. Even when doing what is right may require sacrificing personal allegiances and friendships for professional duty and ethics—for personal honor.

Those are the moments that will truly test the leader within you—test whether you will take the hard path or the easy path, the wrong path or the right path. Always remember, as a wise man once said, "following the path of least resistance is what makes men and rivers crooked."

The willingness always to take the right path, even if it is the hard path, is called character. In every aspect of your life, whether personal or professional, you must always maintain the courage of your convictions—your personal integrity. President John Adams wrote to one of his sons: "A young man should weigh well his plans. Integrity should be preserved in all events, as essential to his happiness, through every stage of his existence. His first maxim should be to place his honor out of reach of all men."

And, I would add, don't kid yourself. More often than not, doing this involves traveling a difficult and lonely road.

In today's age, it may sometimes seem as if ideas like character and honor and patriotism are ideas whose time has passed.

A Scottish general recently told the Pentagon press corps, "I still believe in duty, service, and sacrifice." But, he added that "may be a bit old-fashioned."

By your decision at a young age to choose the path of service, you have already proven that the time for these ideas has not passed—that they are not old-fashioned—that in fact right now, with all the challenges we face, this is the time when we most need to recommit ourselves to them.

The values of duty and service and sacrifice may be old, but that is because they are timeless. They may be obvious, but that is because they are true. They are interwoven into the very fabric of our nation's past, present, and future.

Almost 100 years ago, President Theodore Roosevelt delivered an extraordinary speech called "Citizenship in a Republic." In an oft-cited passage, Roosevelt said:

"In the long run, [our society's] success or failure will be conditioned upon the way in which the average man, the average woman, does his or her duty… The average citizen must be a good citizen if our republics are to succeed."

But Roosevelt went on to say something else that's not often quoted. He added that "the average cannot be kept high unless the standard of the leaders is very much higher."

You are not average citizens—and so you can never be content to be merely "good citizens." You must be great citizens. In everything you do, you must always make sure that you are living up to the highest personal and professional standards of duty, service, and honor—the values of the Air Force, the values of the American armed forces, indeed the values of the United States.

And when you are called to lead, when you are called to stand in defense of your country in faraway lands, you must hold your values and your honor close to your heart. You must remember that the true measure of leadership is not how you react in times of peace or times without peril. The true measure of leadership is how you react when the wind leaves your sails, when the tide turns against you. If at those times you hold true to your standards, then you will always succeed. If only in knowing you stayed true and honorable.

A final point. Today you will take an oath to protect and defend the Constitution of the United States. I have taken that oath seven times over the last forty years—the first when I enlisted in 1966 and the last when I became Secretary of Defense.

Today, as I did with your Naval Academy colleagues last Friday, I want to encourage you always to remember the importance of two pillars of our freedom under the Constitution—the Congress and the press. Both surely try our patience from time to time, but they are the surest guarantors of the liberty of the American people.

The Congress is a co-equal branch of government that under the Constitution raises armies and provides for navies and air forces. Members of both parties now serving in Congress have long been strong supporters of the Department of Defense, and of our men and women in uniform.

As officers, you will have a responsibility to communicate to those below you that the American military must be non-political and recognize the obligation we owe the Congress to be honest and true in our reporting to them. Especially when it involves admitting mistakes and problems.

The same is true with the press, in my view a critically important guarantor of our freedom. When it identifies a problem, as at Walter Reed, the response of senior leaders should be to find out if the allegations are true—as they were at Walter Reed—and if so, say so, and then act to remedy the problem. If untrue, then be able to document that fact. The press is not the enemy, and to treat it as such is self-defeating.

As the Founding Fathers wisely understood, the Congress and a free press, as with a non-political military, assure a free country. A point underscored by a French observer writing about George Washington in 1782. He wrote, "This is the seventh year that he has commanded the army and that he has obeyed the Congress; more need not be said."

Those of us in the chain of command, as well as all of your fellow citizens, are proud of everything the Class of 2007 has accomplished. We are proud of the courage you have displayed by stepping forward to volunteer. We are proud of the courage you have displayed by making it through four years at the Academy. And we are proud of the courage you will display tomorrow as you go out to face great and daunting challenges, and as you face them with unwavering resolve.

We expect great things from you in the years to come. The safety of the nation is in your hands—and there is nowhere else the American people would rather it be.

Congratulations, may God watch over all of you, and these United States.

## International Institute for Strategic Studies

*Singapore, Saturday, June 02, 2007*

THANK you, John.

It is a special pleasure to be in Singapore to attend this conference, the premier forum for exchanging views on security in the Asia Pacific region in a setting aptly named Shangri-La. As one of the most prosperous and stable nations in the world, Singapore has emerged as a key contributor to security in the region, a strategic partner of the United States, and a valued friend to most of the nations represented here today.

This morning I would like to offer some thoughts on key issues relating to Asian security to include the importance of Central Asia while providing some broader historical perspective to inform our discussion.

From its inception as a young republic, the United States has been a Pacific nation. Over the past century we have paid a significant price in blood and treasure to fight aggression, deter potential adversaries, extend freedom, and maintain the peace and prosperity of this part of the world. We have strong interests in all points of the Asian compass East, South, Southeast, Northeast and Central spanning the entire spectrum of economic, political and security relations. Our engagement in Asia has been central to Americas approach to global security for many decades through multiple administrations of both political parties. It remains no less so today, and will become increasingly so in the decades to come.

Some people have suggested that the United States maybe neglecting Asia, because we have been too focused on Iraq, Afghanistan, and other hot spots. In reality, far from neglecting Asia, the U.S. is more engaged than ever before. We have been extraordinarily busy in recent years as we reshape and strengthen our security ties based on shared interests.

Some are bilateral relationships that have been formed, renewed, or modernized each with varying types and degrees of cooperation. Others are the regional arrangements that have formed with our support to deal with common challenges.

America's long-standing alliances in Northeast Asia are being transformed to fit the realities of the 21st Century. The Republic of Korea, with a large, modern military and one of the world's strongest economies, is assuming more responsibility for its own defense while the United States reduces its military footprint. We are realigning and repositioning our forces in Japan while cooperating in new areas such as missile defense. These moves towards achieving a more balanced security partnership should be interpreted as strengthening America's commitment to the defense of two of our closest and long- standing allies. Indeed, in carefully calibrating and refining each of these important relationships, we are guided by one overarching principle: to make each relationship more relevant, more resilient, more responsive, and more enduring.

India and the United States shared an uneven co-existence for much of the Cold War, but since the 2005 summit between President Bush and Indian Prime Minister Singh, we have expanded ties with India the worlds largest democracy and a developing global leader.

Formerly communist Mongolia, has, despite its small population, supported United Nations missions in Africa, and has completed several troop rotations to Afghanistan and Iraq. There is the welcome reestablishment of military to military relations with Indonesia and Pakistan after they were cut off in the late 1990s. Both of these nations have vital roles to play in overcoming jihadist terrorism.

In addition to reinvigorating bilateral ties and forging new partnerships, the U.S. has also been active in key regional initiatives in the areas of counterterrorism, non-proliferation, missile defense, maritime security, and crisis response.

Terrorist attacks in Bali and the Philippines as well as other plots that were foiled including right here in Singapore have made it clear that we need to work together to counter violent extremist networks. The U.S. works closely with regional militaries, law enforcement agencies, and intelligence services to share information, provide training, and in some cases conduct joint operations.

The proliferation of dangerous weapons and materials to terrorists or others is another major threat. The United States knows that we cannot prevent unilaterally the flow of these weapons, which is why we work with partners to improve physical security, interdict shipments, and employ sanctions when necessary. The Proliferation Security Initiative is the cornerstone of this effort and it is showing results in Asia.

Maintaining a reliable deterrent against attacks on America and our allies from ballistic missiles is a critical objective of our national security strategy. As a defensive measure, we are working together to create a network that both deters aggression and provides protection in the case of a missile launch.

We also recognize the importance of maintaining free and secure maritime routes and infrastructure. The Container Security Initiative is one element of our strategy, but President Bush also approved the 2005 National Strategy for Maritime Security to help prevent hostile or illegal acts within the maritime domain. The U.S. Navy and Coast Guard are working with many other countries to develop joint strategies for dealing with both littoral and strategic waterways.

Calamities such as the 2004 tsunami, the devastating earthquake in Central Asia, and the typhoons that ravage this part of the world demonstrate the need for a regional approach to humanitarian crises. The U.S. is committed to assisting Asian nations in their time of need, and we are working with partners to prepare and fine-tune our collective response before disaster strikes. We are also preparing for other non-traditional security threats that result from medical crises such as pandemics, avian influenza, and SARS.

Based on this record it should be clear that the U.S. is not neglecting Asia, and will not do so in the future. We are an Asian power with significant and long term political, economic and security interests.

Our commitments elsewhere notwithstanding, we will fulfill our commitments in Asia.

What takes place at forums such as the Shangri-La Dialogue the nations present, the issues discussed, the relationships strengthened is but one indicator of the changes that have taken place in Asia since I was last in government just over 14 years ago.

At that time, we were still grappling with the implications of the collapse of the Soviet Union, a welcome but cataclysmic event that fundamentally rearranged the world scene in ways we could not predict. The major military threat to peace had receded dramatically, and more and more people around the world were demanding and making progress toward free economic and political systems. In the United States, we were already beginning to pare down large elements of our military, diplomatic, and intelligence capabilities in what some people called a peace dividend.

But even back then, there were early warning signs of what the future had in store. Consider what was happening in 1993, the year I retired as Director of Central Intelligence to enter private life:

- The former Yugoslavia continued to rupture along nationalist and religious lines that had been papered over since well before World War I;
- Russia struggled to overcome corruption, gangsterism, and the economic, political and moral damage done by more than seven decades of communist rule;
- Hopes that Saddam Hussein might be forced from power after his humiliating defeat in Kuwait gave way to the reality that his regime in Iraq would remain for the foreseeable future;

- Afghanistan, having been abandoned by the U.S. and others following the expulsion of the Soviets, was collapsing into anarchy and would, before long, be taken over by an extremist religious sect;
- North Korea announced its intention to withdraw from the nuclear nonproliferation treaty, triggering international concern; and
- In that year, terrorists linked to Al Qaeda bombed the World Trade Center in New York in a failed attempt to bring down its twin towers a foreshadowing of numerous planned attacks against us at home and abroad.

With all this and more, it should have been clear even back then that history, in all of its turbulent dimensions, had not come to an end. It simply had been frozen, and ancient hatreds would thaw with a vengeance and new challenges to peace would appear over the next decade.

The threats that subsequently emerged, the horrific attacks that have taken place, the nations destabilized, and the countless innocents killed have powerfully illustrated the need for civilized nations to come together in new and dynamic ways as we are doing in Asia. In particular, the challenge posed by terrorists inspired by radical ideologies cannot be overcome by any one nation no matter how wealthy or powerful.

It is a campaign being waged by extremists on a global scale, in an arc that extends from the alienated Muslim populations of Europe, to North Africa and the Middle East, to Central and Southeast Asia. Terrorists and their followers have become adept at taking advantage of the ungoverned spaces of the real and virtual worlds to organize, train, and recruit. They have learned to use the strengths of modern and modernizing societies our technology and infrastructure, and in the case of democracies, our freedoms and openness as well.

They are not necessarily seeking to take over countries, as was the case with aggressor states and guerrilla movements of the 20th Century. Instead, groups such as Al Qaeda, along with regional movements such as Abu Sayyaf, Jemaah Islamiyah, and the Moro Islamic Liberation Front in Southeast Asia seek to weaken states, in order to take advantage of lawlessness and vacuums of authority that may result.

We have learned the hard way that allowing failed states to turn into terrorist sanctuaries has catastrophic consequences. The most searing example of this was in Afghanistan, where Al Qaeda planned and operated with impunity for years before striking on September 11. Today, a broad coalition of 42 nations is working together to see that this fate does not again befall Afghanistan.

At the Munich security conference in February, I emphasized to our European allies that the success achieved in Afghanistan so far should not be allowed to slip away because of a lack of commitment or resolve what I called a potential mark of shame for the worlds most powerful military alliance and its partners in Europe and Asia. This responsibility extends to the nations of Asia, and some have stepped forward:

- At nearly a billion and a half U.S. dollars, Japan is Afghanistan's third largest bilateral donor and is building a portion of the Ring Road that will be key to that country's future;
- India is one of the largest financial contributors to this effort, providing assistance in several areas, to include the construction of the new parliament building in Kabul;
- South Korea provides assistance through a health center, reconstruction projects, and a training institute for public officials;
- Australia's Reconstruction Task Force has made a difference by improving local infrastructure, commerce, and security; and
- New Zealand has assumed the lead of a Provincial Reconstruction Team thousands of miles from its shores.

I would urge others to step forward with assistance to Afghanistan in the areas of governance, reconstruction, and counter-narcotics. It is clear that Afghanistan and its newly independent neighbors in Central Asia face steep obstacles as they strive to make the transition into prosperous, secure and fully sovereign nations. At this point I would like to challenge our allies, friends, and partners in the region to do more to help Central Asia in several key areas:

- First and foremost is economic development. Within this broad category, we see infrastructure development whether road and rail, telecommunications, or electricity generation and distribution capacity as the critical enabler and must have. A vibrant economy will provide the people of Central Asia with more opportunities and the terrorists with fewer potential recruits.
- Second is regional integration. Historically, Central Asian states were tied to the Soviet Union, but they were not necessarily linked to each other. Now, the newly independent states have the opportunity to benefit from the mutual support that comes from linking with each other and with the rest of Asia particularly in terms of infrastructure connections.
- Third, is counterterrorism, which cannot solely be addressed in the context of Afghanistan. The entire region is susceptible to the rise of extremist movements so the rest of Asia has a large stake in making sure Central Asian nations are equipped to deal with this threat.
- Fourth, capacity building initiatives, such as providing advisors to ministries, can promote political and economic reforms. Weak government institutions lead to internal instability. Southeast and East Asia governments are in a unique position of being able to offer their own lessons learned from recent experiences in setting up independent and responsive governments.

- Fifth, Central Asian states need help with counter-narcotics. Asian governments have grown increasingly sophisticated in their counter-drug operations and this knowledge can be transferred.
- Finally, an area in which I am personally focused is security assistance. The United States is promoting defense reform to establish capable, civilian-controlled and responsible defense establishments. Our goal is to help them assert their sovereignty and independence so they are better able to resist external political pressures from neighbor states. I believe that Asian states can offer more military trainers, peacekeepers and advisors to further this effort.

In addition, the U.S. recognizes that some degree of integration of Central Asian states into the broader Asian community would be helpful. Other Asian states can assist in promoting peace and stability in these newly independent states by more actively welcoming the new states into the Asian security structure. Our efforts can balance initiatives being promoted by Russia and China in the region. These efforts need not be competitive, but rather complementary.

Integrating these newly independent states into the fold of the greater Asian family is in the interest of every country represented in this room. Of course, the degree that Central Asian states and Afghanistan choose to integrate into greater Asia is a decision for each of those sovereign countries. We will not presume to make those decisions for them, but it is important that the welcome mat be out for them. The failure to do so could ultimately have devastating results.

In the Middle East, Al Qaeda has sought to turn large swaths of Western Iraq into a base of training and operations. In recent months, the United States has reaffirmed its commitment to protect our allies and long-standing interests in the Gulf region. A new military strategy is in place a strategy focused on providing basic security to the Iraqi people. It is being bolstered by a new emphasis in the political, economic, and governance areas designed to improve the quality of life for all Iraqis. The immediate goal is to create the breathing room necessary to allow reform and reconciliation to go forward steps that will give all of Iraq's communities, majority and minorities alike, a stake in that nations future. Whatever your views on how we got to this point in Iraq, it is clear that a failed state in that part of the world would destabilize the region and embolden violent extremists elsewhere. The effects of chaos in either Central or Southwest Asia will not recognize national, continental, or regional boundaries.

In many ways the security challenges I've described today are unprecedented. But history can still be instructive. As an old Cold Warrior, I believe that many of the principles that guided our response to that epic, multi-generational conflict are relevant today.

As was true in the Cold War, overcoming violent extremists will require a long, sustained effort measured in decades rather than years. It is an asymmetric conflict characterized by insurgencies of various sizes and durations. It is fundamentally an ideological struggle, where the appeal of principle and the power of example provided by secure, prosperous, and tolerant societies will become the decisive edge. It will take strong alliances and vibrant coalitions.

And as with the Cold War, there will be disappointments and divisions and setbacks.

Consider that the Korean War ended in frustrating stalemate during the early 1950s. But the initial stand taken by the United Nations, and the commitment that followed, allowed South Korea to develop into an economic powerhouse and a vibrant, free society that is now taking substantial responsibility for its own defense while reaching out to assist others. The war in Vietnam was a wrenching, costly, and tragic experience for the United States and the Vietnamese people. But by the time of the fall of Saigon in 1975, several other nations of the region had been fortified to withstand communist takeover; there was rapprochement between the United States and China; and though we didn't know it at the time, trends were in place that would ultimately lead to the fall of the Berlin Wall just over a decade later. Today, Vietnam has joined its neighbors as a member of ASEAN, is integrated into the world economy, and has established military-to-military relations with the United States.

Despite the setbacks and frustrations, the coalition opposed to Soviet domination eventually succeeded because over time they got the big things right, and were resolute and united when it counted. Within the United States, administrations of both political parties embraced an unwritten consensus—that communist expansion needed to be contained a consensus that extended, despite many differences and disputes over the years, to our allies and partners in Europe and Asia.

Like the radical movements and insurgencies of the past, there is no doubt that the extremists have their share of advantages that are only multiplied by new information and communications technologies. Among these is the ability to operate as a network, with all its flexibility and resilience. I would argue that in Asia we have a network that is to our advantage as well a network of cooperation and dynamic partnerships that I described earlier between nations of shared interests.

These strong partnerships are the source of our strength as we confront a range of challenges with global implications. One serious challenge is the production and proliferation of dangerous weapons by dangerous regimes—a common threat requiring a concerted and clear-eyed response.

In Northeast Asia, the Six-Party process had a stabilizing effect after North Korea's nuclear test last year, helping to head off a more dangerous reaction from other nations. This process ultimately led to the breakthrough agreement in September 2005, and, in this past February, a complementary agreement on initial steps in the de-nuclearization process. Although the North Koreans have not yet followed through on their commitments, the fact remains that agreements are now in place that will permit resolution of the issue. The joint, vigilant efforts by nations with aligned interests on this issue have given the people of the region a sense that a mechanism is in place to deal with the long-standing problem of North Korea's behavior and ambitions.

Iran poses a similar threat to Southwest Asia and potentially to Europe, particularly as it builds missiles of increasing range. The U.N. Security Council has come together on several occasions—most recently with a unanimous resolution in March to hold Iran to account for its violations of the nonproliferation treaty.

The nuclear and ballistic missile programs of nations like Iran and North Korea pose one set of problems for their neighbors. Another, equally worrisome, possibility is that regimes may sell these weapons and materials to others, including terrorist organizations.

The cooperation needed to overcome these and other of this Century's most vexing challenges will require a significant level of trust and transparency between nations that may have differing perspectives and histories, but should nonetheless share the same basic security concerns. Conversely, distrust and secrecy can lead to miscalculation and unnecessary confrontation.

Russia, for example, has concerns over the prospect of a ballistic missile defense site in Europe and how that system would affect Russia's nuclear deterrent. In April, I traveled to Moscow and proposed to President Putin an unprecedented partnership with Russia in missile defense, and invited Russian officials to inspect our existing interceptor site in Alaska and radar in California.

This is all very real. Not rhetoric and not gamesmanship. The key to turning this potential disagreement into a real opportunity is openness and cooperation. As I said in Munich, one Cold War was enough though there are speechwriters in Moscow who seem to yearn for the old days of bluster and confrontation.

The United States shares common interests with China on issues like terrorism, counter proliferation, and energy security. But we are concerned about the opaqueness of Beijing's military spending and modernization programs—issues described in the annual report on the Chinese armed forces recently released by the U.S. government. But as General Pete Pace, our Chairman of the Joint Chiefs of Staff, pointed out, there is some difference between capacity and intent. And I believe there is reason to be optimistic about the U.S.-China relationship.

We have increased military-to-military contacts between all levels of our militaries, most recently dramatized when General Pace sat in the cockpit of the top-of-the-line Chinese fighter during his last visit. We obviously have a huge economic and trade relationship. Indeed, I have been told that if just one American company, Wal-Mart, was a country, it would be China's eighth largest trading partner. The second meeting of the U.S.-China Strategic Economic Dialogue concluded last week in Washington, D.C., a process designed to improve our economic bilateral relationship. As we gain experience in dealing with each other, relationships can be forged that will build trust over time.

I opened these remarks by noting that the term Shangri-La was an apt description for this conference's setting the luxurious hotel, the beautiful city. The term actually comes from a novel written some 70 years ago. The book described a valley high in the mountains of Tibet, where the inhabitants were blessed by long life and suffered none of the afflictions of the modern world. The residents of that fictional Shangri-La were convinced that the rest of the world would soon destroy itself, reflecting the fears of the author's generation during the 1930s. In a few short years, the calamity of World War II and the nuclear standoff that followed almost proved those fears correct.

But history took a different path. Not towards paradise or tranquility, but where, through commitment, partnership, cooperation, and resolve we are overcoming current challenges to our freedom, prosperity, and security as we overcame the threats of the past.

Thank you. I look forward to responding to your questions.

## Commemoration of the 63rd Anniversary of D-Day

*Colleville-Sur-Mer, France, Wednesday, June 06, 2007*

SIXTY-THREE years ago, the vanguard from the free world landed on these beaches—forces from the United States, England, Canada, Australia, France, and other nations. They had a singular purpose: to destroy the entrenched forces of oppression—what Churchill called "the foulest and most soul-destroying tyranny"—so that this nation, this continent, and this world could one day know the tidings of peace.

Presidents and statesmen have paid homage here. Yet, no matter how moving the words, how powerful the speaker, all were humbled to stand in front of those men who sacrificed on these shores their safety, their innocence, and often their lives.

On June 5th, 1944, a mass of men and ships the likes of which the world had never seen set sail from across the Channel. An intelligence officer later described the awesome sight. "The vast machinery of invasion had started to move, inevitably and relentlessly. It was exhilarating, glorious, and heartbreaking."

For those who were here, the next day, June 6th, unfolded as if it were a lifetime. Men who had only recently felt the warmth of their families now felt the frigid waters of the English Channel and the lonely sands of a war-torn, windswept beachhead. Men who had just a few months earlier been boys in the midst of adolescence suddenly found themselves traversing a warren of lethal obstacles on beaches named Omaha, Utah, Gold, Juno, and Sword.

Here, at what came to be known as "Bloody Omaha," the allied offensive faltered and almost failed. Bad intelligence, bad weather, bad luck—they all conspired against the landing forces.

In its horror, the scene was breathtaking. One GI said that the first men on the beach tumbled "just like corncobs off of a conveyor belt." Another described Omaha as nothing more than "the dead and dying and the people coming in to replace them." That day, the blood of the dead and the dying turned the seas red.

Even so, stories of valor were countless. As gunfire rained down, men stopped to pull comrades from the water. Alone or outnumbered, they charged heavily fortified positions.

We heard this morning from Medal of Honor winner Walter Ehlers, who after suffering wounds himself, would carry one of his riflemen to safety, refused medical evacuation, and returned to lead his squad as the spearhead of the attack.

Lieutenant Colonel James Earl Rudder—one of my predecessors as president of Texas A&M University—was struck twice while leading his Rangers up and over the cliffs of Point du Hoc. Others wounded continued to fight on as they took their very last breath.

No amount of firepower could overwhelm their instincts, their bravery, their compassion, and their humor. Captain Frank Corder of Texas stepped onto the beach, and, as bullets and bombs whizzed by, said, "This is no place for Mrs. Corder's little boy Frank."

Ahead of Mrs. Corder's little boy and all the troops pushing inland still lay hundreds of thousands of determined enemies ready to fight in the hedgerows and apple orchards of Normandy, in the forests of the Ardennes, and finally in the narrow streets of German villages.

But on these beaches the tide finally turned—though at a great and sorrowful cost. This cemetery is the final resting place for 9,387 Americans and a memorial to 1,557 who went missing during the Normandy campaign. All are remembered and honored here. All gave the last full measure of devotion.

We mourn every man who fell, even as we quietly give thanks for their sacrifice. We are grateful that those who survived these beaches and subsequent battles one day laid down the weapons of war and pursued the dreams of peace. And we are grateful that out of the rubble of war free nations conceived of and built a better future.

Today we mark another chapter at this hallowed place with the opening of a new visitor center.

We build memorials like this to remind us of the past. So that successive generations will know the enormous cost of freedom. So that our children and grandchildren will never forget the stories of those who fought here. So that the passage of time and the thinning of their ranks will never dim the glory of their deeds.

Minister Morin, events like this also remind us of all we have endured together—remind us of our long history in times of war, and in times of peace—remind us of the shared values that transcend whatever differences we may have had in the past, or may have in the present.

Sixty-three years ago, we fought together in the belief that the blood of free men could wash away the stain of tyranny. Some thought that we were perhaps overly ambitious—and indeed we were immediately challenged by another soul-crushing tyranny that built new walls of oppression and covered much of this continent in a new darkness. But again, the free nations of the world came together to defend their civilization and their hard-earned freedom. Throughout the many years of that long, twilight struggle, our partnership had its share of divisions and discord. But even when we disagreed on tactics, we remained unified in purpose. And today, finally, the dream of a Europe whole and free is a reality.

It is another challenging period for the nations of the free world. We once again face enemies seeking to destroy our way of life, and we are once again engaged in an ideological struggle that may not find resolution for many years or even decades.

At the same time, many people believe that the foundations of the alliance forged in places like this have collapsed or outlived their usefulness.

Let the people of our nations never forget that we are bound by history and values just as we are bound by blood. The blood of Americans. The blood of Frenchmen. The blood of Britons. The blood of other allies. The blood of everyone who has ever perished in defense of the lofty ideals that gave rise to and still underpin our great alliance. Those ideals were given birth on this continent, and given their renewal on battlefields like this one.

Today we are reminded of the frailty of human life and the terrible cost of war. But we are also reminded of the spirit of friendship and the strength of unity that has sustained the hope of both our nations, and all the nations of the free world, during many dark days.

I leave the last words of this ceremony to a man who fought on these shores—Captain Joseph Dawson, one of the first officers to make it beyond the bluffs of Omaha beach. In a letter to his family shortly after D-Day, he wrote:

"[H]ere amid the apple trees of this bit of France, with the symphony of war encompassing me, I have found peace of heart and soul never before attained in all my life, for here I am with the bravest, finest, grandest bunch of men that God ever breathed life into. Before it's all over, you will know that this is true and that this company is my life."

May God never let us forget what happened here. And may God grant peace of heart and soul to everyone who fought on these shores. Thank you.

## U.S. Special Operations Command Change of Command Ceremony

*Tampa Convention Center, Tampa, Florida, Monday, July 09, 2007*

GOOD morning. It's good to be back in Tampa. I was here last March for the CENTCOM change of command down at MacDill, and I see some of the same faces:
- General Pace. Thank you for you being here. When the Marines joined Special Operations Command for the first time, it garnered quite a lot of attention. As the first Marine Corps Chairman of the Joint Chiefs of Staff, General Pace clearly met the high standard required for joining the roster of Marine Corps historic firsts. I am grateful for his advice and his leadership, and I thank him for his long and dedicated service to our country;
- General Casey, General Schwartz, Admiral Fallon, Admiral Allen, it's good to see you all;
- I also want to thank Congressman Bill Young for being here today; and
- Two former SOCOM commanders, Generals Schoomaker and Holland are here. Welcome.

I understand that when this command was first considered, the plan was to base it in Washington, D.C.

The first commander, General Lindsay resisted, noting that, "I didn't want SOCOM to become another staff agency." Knowing the kind of men and women who join special ops, I think there was little danger of that. But I am certainly grateful for the decision to base it here, as, I am confident are the personnel of this organization.

My thanks to the folks from the Tampa community for all that you do to support SOCOM and CENTCOM—two commands that are critical to our nation's campaign against violent extremism.

Above all, I want to recognize the men and women of SOCOM, and the special operators throughout the world. Your task is anything but easy. You have volunteered multiple times to take the most difficult assignments. You do so with courage, determination, and skill that leaves the rest of us in awe.

The success of special operations begins with the individual warrior—each one of you—and we are eternally grateful for your willingness to serve our nation.

I also want to congratulate this command on 20 years of service and success. Special Operations Command—the role of special operations in general—has certainly come a long way.

Personally, I am still haunted by Operation Eagle Claw—the mission in 1980 to rescue hostages in Iran—a disaster that provided a painful, but ultimately valuable, glimpse into the shortfalls that existed at the time. I was then the executive assistant to the Director of Central Intelligence. I spent a very long afternoon—and an even longer night—with the Director and other officials at the White House as news of what had happened in the Iranian desert trickled in. We mourned the deaths of the servicemen. We knew the tremendous abilities of those involved—men like Pete Schoomaker who would go on to lead this command and become Army Chief of Staff. We were convinced the mission could have succeeded. But the failure of that mission showed a number of serious weaknesses, to include the ability of different units and services of the special operations world to work together.

So like we always seem to do in this country, we studied our mistakes and learned. Joint capabilities would eventually eclipse parochial service interests. And this year, we celebrate 20 years of service from a command that is at the forefront of today's fight to preserve our freedom and our way of life. U.S. special operations forces are engaged in missions as varied as strengthening America's partners in Southeast Asia to hunting Al Qaeda in the most dangerous regions of Afghanistan and Iraq. Our country no longer needs to search for unseen benefits provided by your unique talents and capabilities. We experience them every day. We are safer for it, and grateful for it.

Today we recognize a man who for nearly four years has been leading them in this fight.

General Doug Brown's career, in some ways, reflects the dynamics that created this command. Shortly after the Iran mission, then-Major Brown joined the new Special Operations Command at Fort Bragg. He served with Task Force 160, learning from the difficulties experienced during Operation Urgent Fury in Grenada, and later led a battalion of Nightstalkers during Operation Desert Storm.

He came to this post four years ago determined to improve the way special operators fight. He has done just that.

But he also improved the way this command works. One of General Brown's first major undertakings as Commander was the vital reorganization of elements in the Center for Special Operations. His "course correction" allows different specialties to build on each other rather than compete. After Desert One, there was lingering bitterness and ill will over the role of—or lack of—intelligence during the operation. Today, intelligence officers and special operators work side-by-side—fusing their expertise and planning to greatly improve results.

General Brown also strengthened the way our forces work with units from other countries—emphasizing language and culture training to build trust and bonds with foreign militaries. And his emphasis on indirect operations aimed to prevent minor problems from growing into much bigger crises.

Special operators will always be able to respond with strength and force when the mission calls. But as they work side-by-side with people throughout the world, members of this command have demonstrated our nation's values, as well as fostered cultures and communities that are more likely to reject extremism and violence.

Today, there is little that is beyond the capabilities—or the reach—of special operators. This is due in no small part to your leadership, General. Thank you for all that you have done in four decades of service for our country.

Of course, even with a long and successful career, Doug might say that his biggest accomplishment in life was marrying his high school sweetheart. Penny, this would usually be the point where I congratulate you for your decades of service to our nation and wish you well. That I certainly do. But with two sons-in-law in the Army, and at least one grandson your husband once described as a "Green Beret in training," you still have your hands full. Thank you for all that you have done for Doug, your family, this command, and for our country.

General Brown's successor, Admiral Eric Olson, is a true warrior and legend in this community.

In 1993, after a Black Hawk helicopter was shot down during an operation in Somalia, hundreds of enemy fighters surrounded the crash site and the stranded pilot.

No doubt most of you know of the ensuing actions of Congressional Medal of Honor recipients, Sergeants Shughart and Gordon, who volunteered to secure the crash site and gave their lives to save the downed helicopter pilot. And you undoubtedly recognize the creed driving those Sergeants' actions: Never leave a comrade behind. For those who know Admiral Eric Olson, you also know that he more than believes in that creed—he lives it. For it was then-Commander Olson who fought street-by-street, leading a ground convoy to rescue his comrades after the Black Hawk went down.

As Deputy Commander here, Admiral Olson has worked tirelessly to leave no special operator behind—in battle, in training, in quality of life, and in any aspect where he could make a difference.

As the first Navy SEAL to wear three stars, and now four, there is no mistaking his combination of courage, experience, and leadership.

When Congress was debating how to modernize this organization, the commander at the time, Major General Scholtes, went to the extraordinary step of first retiring to ensure he could give his most honest and unvarnished advice when testifying about the proposed changes.

Eric, I expect you'll stick around. But I also hope that you continue your custom of giving honest opinions and recommendations—with the bark off, and straight from the shoulder. The men and women here today, and around the world, already benefit from your dedication to them and to our nation. And I look forward to working with you as our nation continues operations across the globe, and I wish you and Marilyn all the best.

# Marine Corps Association Annual Dinner

*Crystal Gateway Marriott Hotel, Arlington, VA, Wednesday, July 18, 2007*

THANK you, General Palm, for that kind introduction. And let me express my appreciation to the Marine Corps Association for inviting me to speak at your first annual dinner. I'm honored to be in the presence of so many who have devoted their lives to the defense of this country.

Senator Warner, General Magnus, it's good to see you. I must say that Senator Warner is a special friend. He's introduced me to the Senate for confirmation four times. If it's not a record, it has to be close to a record. And I have to say the first time was more than 20 years ago. That dates us both, but I would have to say I think he is a special friend and a great American. So thank you for being here tonight.Well, I can't tell you what a pleasure it is to be back in Washington again. A place where, as Senator Alan Simpson used to say, "those who travel the high road of humility encounter little heavy traffic." Or as others would say, "where there are so many lost in thought because it's such unfamiliar territory." Where people say I'll double-cross that bridge when I get to it. The only place in the world you can see a prominent person walking down lover's lane holding his own hand. They say Washington's a city of monuments. I have to say the most monumental things that I've seen in over 40 years is the egos of some of the people I've worked for, and I have to tell you the most monumental ego was the first president I worked for, Lyndon Johnson.

Johnson once had the chancellor of the Federal Republic of Germany, Ludwig Erhard, to the LBJ ranch, and Erhard at one point said, "Well, Mr. President, were you born in a log cabin?" And LBJ responded, "Why no, Mr. Chancellor, I was born in a manger."

Or the time he gave a stag dinner in the White House and Bill Moyers was there and Moyers was a White House staffer seated below the salt, where White House staffers belong. And Johnson asked Moyers to ask the blessing and Moyers started to pray and a few seconds into the prayer, Johnson lifted his said, looked down at Moyers and said, "Bill, I can't hear you." And Moyers, without lifting his head, looked and said to the president, "That's cause I'm not speaking to you."

It's also a city of monumental embarrassments. Like the first time that President Nixon met with Israeli Prime Minister Golda Meir. He had just appointed Henry Kissinger as Secretary of State. And Golda Meir had with her her foreign minister, Abba Eban, who had a doctorate from Oxford University. And Nixon turned to Golda Meir and said "Just think, Madam Prime Minister, we now both have Jewish foreign ministers." And Golda Meir looked at him and said, "Yes, but mine speaks English."

But I think the most embarrassing moment during my career was when Nixon visited Italy and he met with the Pope, and Melvin Laird was along as Secretary of Defense. Kissinger and Nixon decided that Laird shouldn't be invited to the meeting with the Pope, as sort of the Minister of War.

And so, Nixon was in the next morning having his private audience with the Pope, and the rest of us were waiting outside. And who should come striding down the hall smoking an enormous cigar but Laird. He had clearly found out about the meeting, probably through good military intelligence.

And Kissinger was kind of beside himself, but he finally said "Well, Mel, at least extinguish the cigar." So Laird stubbed out his cigar and put it in his pocket.

The American party a few minutes later went in to their general meeting with the pope. Pope was seated at a little table in front, Americans in two rows of high-backed chairs. Back row, Kissinger on the end; Laird next to him. A couple of minutes into the Pope's remarks, Kissinger heard this little patting sound, and he looked over, and there was a wisp of smoke coming out of Laird's pocket. The Secretary of State thought nothing of it. A couple of other minutes went by and the secretary heard this patting sound, slapping going on, and he looked over and smoke was billowing out of Laird's pocket. The Secretary of Defense was on fire.

The American party heard this slapping, and thought they were being queued to applaud. And so they did.

And Henry later told us, "God only knows what his Holiness thought, seeing the American secretary of defense immolating himself, and the entire American party applauding the fact."

Well, it's hard to believe that it was only seven months ago—to the day, as a matter of fact—that I began my current job. And as many of you know, navigating the Pentagon can be quite a challenge.

Newsman David Brinkley used to tell the story of the early days at the Pentagon, when woman told a Pentagon guard she was in labor and needed help quickly in getting to a hospital. The guard said, "Madam, you should not have come in here in that condition." And she answered, "I wasn't in that condition when I came in here."

One of the main reasons I have managed to get around—and get by— at the Department these past seven months, is a great officer who has the distinction of being the first Marine Corps Chairman of the Joint Chiefs of Staff. This summer, General Peter Pace marked 40 extraordinary years of active service. Pete's traveling overseas this evening or he would be here tonight. In fact, he told me that I should accept the invitation to this event.

I'm sure most, if not all of you, are unhappy that Pete will not continue on for a second term as chairman. I am as well. Pete Pace has been my friend, my partner, and my mentor. I trust him completely; I value his candor; and, I enjoy his sense of humor.

I told Pete several months ago that it was my intention that we work together until I left on January 20, 2009. I can't tell you how much I regret that the current environment here in Washington did not make that possible. I am deeply grateful for Pete's 40 years of devoted service to our country, a sentiment I am confident is shared by you and by all Americans.

Last month, after visiting with U.S. and NATO forces in Afghanistan, I had a chance to speak at Normandy on the 63rd anniversary of D-Day. It was a powerful experience to stand among those crosses—thousands of them, row upon row—and reflect on the magnitude of what had been accomplished on that day—and at what cost.

The story behind how America developed the means to put men on those beaches is, I think, instructive. In the late 1930s, the Marine Corps was still grappling with how to move troops from ship to shore under hostile fire. At the time, and after the disastrous Gallipoli campaign of the First World War, such a maneuver was considered foolhardy at best, and suicidal at worst. In 1937, a Marine 1st Lieutenant, Victor Krulak, was stationed in China. And during a Japanese amphibious assault on Shanghai, Krulak borrowed a tugboat to get a better look. He saw—and clandestinely photographed—Japanese men and equipment coming onto the beach from a landing craft with a retractable ramp.

Lieutenant Krulak sent those photos and an accompanying report back to Washington. You can imagine what happened next. They gathered dust in a cabinet, with a note labeling them, and I quote: "the work of some nut in China."

Krulak eventually returned to Washington, and doggedly pursued his idea until a Marine general hooked him up with an eccentric New Orleans boat maker named Higgins. The result, as all of you know, was a landing craft with a retractable ramp that was introduced by the thousands and was used to carry Allied forces to liberate Europe and much of Asia.

Krulak's was, of course, a legendary career: Navy Cross; counterinsurgency advisor to the Joint Staff; commander of the Fleet Marines in the Pacific during the Vietnam War; and, father of a future Marine Commandant, Chuck Krulak, with whom I met yesterday. Victor Krulak's story and accomplishments teach us a good deal:

- About learning from the experiences and setbacks of the past;
- About being open to take ideas and inspiration from wherever they come; and
- About overcoming conventional wisdom and bureaucratic obstacles thrown in one's path.

In the years since September 11th, hundreds of thousands of our troops have done all these things and more in Afghanistan, Iraq, and elsewhere around the globe. There are the Marines who set up a daily news report over loudspeaker—"the Voice of Ramadi"—to counter the hostile propaganda blaring out of some of the mosques. Then there is an Army staff sergeant, a field artillery radar specialist, who was elected a sheik by Iraqi village elders for his work in their communities. He was given white robes, five sheep, and some land; he was advised to take a second wife—a suggestion frowned upon by his spouse back in Florida.

But in these campaigns, the men and women wearing our nation's uniform have assumed the roles of warrior, diplomat, humanitarian, and development expert. They've done so under the unblinking, unforgiving eye of the 24-hour news cycle while confronting an agile and ruthless enemy. And they've done it serving in a military that has for decades been organized, trained, and equipped to fight the "big wars" rather than the small ones. They have shown what General Victor Krulak later wrote was the "adaptability, initiative and improvisation [that] are the true fabric of obedience, the ultimate in soldierly conduct, going further than sheer heroism."

For the next 10 minutes or so, I'd like to offer some thoughts on where our military—and our government—must apply the lessons that we've learned from the ongoing conflicts to build the capabilities we will need in the future. These points are clear:

- Our military must be prepared to undertake the full spectrum of operations—including unconventional or irregular campaigns—for the foreseeable future.
- The non-military instruments of America's national power need to be rebuilt, modernized, and committed to the fight.
- And third, we must think about, envision, and plan for, the world, the future—of 2020 and beyond.

This is necessary because in the decades ahead, the free and civilized world will continue to look to the United States for leadership, despite all of the challenges today. Churchill once said that "the people of the United States cannot escape world responsibility," what he called the "price of greatness." This responsibility calls on us to prepare for threats other than those on our television screen every night, challenges that are on or beyond the horizon.

America's conventional forces—air, land, and sea—will continue to be called on to deter cross-border aggression, protect the sea lanes and energy supplies, and send a message of strength and resolve to friends and potential enemies—be they nation-states or other actors. These formations must move with speed and agility to a range of potential fights, with deployment times measured in weeks or days rather than weeks and months.

Above all, it's clear the United States and our allies will continue to be threatened by violent extremists, almost always operating in countries with whom we are not at war. The ambition of these networks to acquire chemical, biological, and nuclear weapons is real, as is their desire to launch more attacks on our country and on our interests around the world. And as we saw most recently in the United Kingdom, the barrier to entry—in resources and sophistication—remains low when the goal is simply to disrupt or terrorize.

In recent years, America has fully joined the battle in a war that was declared on us a long time ago.

I remember vividly a day in December 1991, when as CIA Director I—along with then-Secretary of Defense Dick Cheney—attended an arrival ceremony at Andrews Air Force Base. We were there to receive the remains of two men—two of our nation's "bravest sons"—who had been kidnapped, tortured, and murdered by terrorists in Lebanon. One was William Buckley, CIA station chief in Beirut. The other was Marine Lieutenant Colonel William Higgins, who served with the U.N. peacekeeping force in Lebanon.

These two Americans were murdered by the same Hezbollah-linked extremists who killed hundreds of Americans in 1983 at the Marine barracks and U.S. embassy in Beirut. It is important to remember that until the morning of September 11, 2001, Hezbollah had been responsible for the deaths of more Americans, our countrymen, than any other terrorist group in the world.

Now we must deal with an even more deadly threat. Since Al Qaeda attacked America nearly six years ago, our armed forces have been tasked with removing hostile regimes and booting out terrorist networks in Iraq and Afghanistan; initially quick military successes that in both cases have led to protracted stability and reconstruction campaigns against brutal and adaptive insurgencies.

And though these conflicts will not last indefinitely in their current form and scale, we must expect our military to be called to other irregular campaigns in the future.

What we now call "asymmetric war" has become a mainstay of the contemporary battlefield, if not its centerpiece. Indeed, after Desert Storm and the initial military success of Operation Iraqi Freedom, it is hard to conceive any country challenging the United States using conventional military ground forces—at least for some years to come.

However, history shows us smaller, irregular forces—insurgents, guerrillas, terrorists—have for centuries found ways to harass and frustrate larger, regular armies and sow chaos.

Today, the "three block war" that Commandant Chuck Krulak predicted in the 1990s—where small units would simultaneously conduct combat, stability, and humanitarian operations in urban landscapes—has become a daily reality for American servicemen and women. In these situations, America's traditional edge in technology, firepower, and logistics provides important tactical advantages, but not the necessary strategic success.

Direct force will no doubt need to be used against our adversaries—ruthlessly and without mercy or apology. But it is also clear that in these kinds of operations, we are not going to kill or capture our way to victory.

Today in Iraq, General Petraeus and Ambassador Crocker are implementing a strategy based on targeting Al Qaeda, co-opting some insurgents and marginalizing others, and providing basic security and an improved quality of life for the Iraqi people. It will take patience and persistence, and some level of American force and assistance, for some time.

Looking forward, tasks such as standing up and mentoring indigenous armies and police—once the province of Special Forces—are now a key mission for the military as a whole. The same is true for mastering foreign language and civil affairs tasks such as reviving public services and promoting good governance. They have moved from the margins to the mainstream of military thinking, planning, and personnel policies, where they must stay. But as much as the armed forces must be prepared to take on these tasks, the fact remains that much of the necessary expertise belongs in other parts of our government.

We're still struggling to overcome the legacy of the 1990s, when so many of the key non-military capabilities in the American government—in diplomacy, strategic communications, international development, and intelligence—were slashed or eliminated following the end of the Cold War.

During the 1990s, the State Department froze new hiring of Foreign Service officers. I was in the White House in the Carter administration after the fall of Iran, and we had a group called the political intelligence working group and we examined what had happened. And among other things, we determined that in 1979, in the embassy in Riyadh, we had two Foreign Service officers who spoke Arabic and they spent 40 percent of their time squiring around CODELs.

The United States Information Agency, which had been an enormously successful organization for communicating America's values and message around the world, was abolished in the 1990s as an independent entity and folded into the State Department—a shadow of its former self. The Agency for International Development saw deep staff cuts—its permanent staff dropping from a high of 15,000 to 3,000 today, becoming essentially an outsourcing and contracting agency.

Today, the total number of U.S. government civilian employees working in the Provincial Reconstruction Teams in both Iraq and Afghanistan is approximately two hundred.

So, the goal for us must be an integrated effort, a reinvigoration of all elements of national power. It will require a serious commitment of resources and priorities from the Congress and the country. I believe we have little choice if we are to secure our nation and our freedoms in the years ahead.

I've spoken tonight about what the Pentagon calls the "non-kinetic" aspects of war. It is a sad reality, however, that throughout human history, some have always thought and sought to dominate others through violence and crimes against the innocent. When all is said and done, they understand and bow not to reason or to negotiation, but only to superior force. Thus we should never lose sight of the ethos that has made the Marine Corps—where "every Marine is a rifleman"—one of America's cherished institutions and one of the world's most feared and respected fighting forces.

I began my remarks this evening with a story about an extraordinary young Marine officer, Victor Krulak; I will close with another.

On one wall of my conference room there is a large, framed photo of a Marine company commander taken during the first battle of Fallujah, in April 2004. He's speaking into a radio handset while giving directions to his men as combat rages just blocks away. It's a shot that could have been taken of any number of Marines in any number of places over the last century—at Tarawa, at Inchon, or of Lieutenant Peter Pace at Hue, in 1968.

During that Fallujah battle, Captain Douglas Zembiec and some men from his Echo Company were on a rooftop drawing rocket propelled grenades from all directions. They tried to radio a tank crew for support, but couldn't get through. Zembiec raced out onto the street through withering fire, climbed onto the tank, and directed the gunner where to shoot.

After the battle, he said that his Marines had "fought like lions," and he was soon himself dubbed "the Lion of Fallujah." He was an unabashed and unashamed warrior, telling one reporter that "killing is not wrong if it's for a purpose, if it's to keep your nation free or to protect your buddy." Zembiec's battalion operations officer described him as someone who "goes out every day and creates menacing dilemmas for the enemy."

A newspaper profile at the time described him as a "balding, gregarious man who, in glasses, looks like a high school science teacher."

After returning from Iraq, Doug was promoted and given a desk job at the Pentagon. He chafed at the assignment, volunteered to deploy again, and was sent back to Iraq earlier this year. This time, he would not return—to his country or to his wife Pamela and his one-year old daughter.

In May, the Lion of Fallujah was laid to rest at Arlington and memorialized at his alma mater in Annapolis. The crowd of more than 1,000 included many enlisted Marines from his beloved Echo Company. An officer there told a reporter: "your men have to follow your orders; they don't have to go to your funeral."

Every evening, I write notes to the families of young Americans like Doug Zembiec. For you, and for me, they are not names on a press release, or numbers updated on a web page. They are our country's sons and daughters. They are in a tradition of service that includes you and your forebears going back to the earliest days of the republic.

God bless you, the Marine Corps, the men and women of our armed forces, and the country we have all sworn to defend.

Thank you.

## Farewell To Admiral Giambastiani

*Annapolis, MD, Friday, July 27, 2007*

WELL, I can't possibly top that. I have no animals in the background. Mr. Vice President, Mrs. Cheney—you honor us by being with us today. Welcome to all the other distinguished guests who are here.

It is difficult, in a short amount of time, to describe all that Admiral Giambastiani has meant to those he's worked with—or to sum up all that he has accomplished in his decades of service.

His story has been a profile in excellence—of steep challenges and enormous achievements. But his successes likely mean less to him than the successes of others. Ed's official biography notes that he is most proud of the 19 unit awards and commendations he has been associated with because, quote, they "recognize the participation and accomplishments of the entire team."

Ed is the type of person who makes combining distinction and humility an art form. Anyone who has sat in a meeting with Ed also knows his good humor.

Ed once joked that the two biggest lies in the Navy occur when an inspection team boards a ship saying "we're here to help," and when the ship's captain answers "welcome aboard." That may be true. But Ed never shied away from criticism or difficult tasks.

In fact, he took on his first challenge right here at the Naval Academy. By the time young Ed Giambastiani arrived on the Yard in 1966, more than 250,000 Americans were fighting in Vietnam. He could have chosen an easier path to earning a bachelor's degree—and certainly an easier path after receiving a degree—but like his father before him, he volunteered to serve in the Navy.

In 1975, the shift to an all-volunteer force was just two years old. The Navy was confronted with problems of discipline, drug use, and prejudice, but Lieutenant Giambastiani took on the challenge of what he called a "transformational event" and became an enlisted program manager at the Navy's recruiting command.

Long considered one of the service's sharpest and most innovative minds, Ed's first command was the Navy's only nuclear powered deep diving research submarine. His "reward" for a successful command tour was coming to work for me at CIA.

By the time of his arrival at the Agency in 1985, the Soviets were intensifying conflicts in Afghanistan, Angola, Nicaragua, and elsewhere, and CIA was heavily engaged in countering the threats. Ed and I were in the middle of a virtual intelligence war. It was a hectic time, to say the least. I don't recall asking if Ed ever longed to return to the relative calm of the deep ocean.

Later, as head of Joint Forces Command, he rightly observed that too often lessons learned are catalogued but not acted upon—a critical deficiency during an asymmetric war in the information age. So Ed worked closely with General Tommy Franks and his successors at Central Command to identify and absorb those lessons with unprecedented speed. He did such a good job that he was chosen to streamline and transform the NATO alliance into a more expeditionary force as well.

Most recently, nearly four years into a global struggle against radical extremists, Ed stepped forward to help lead the fight as vice chairman of the Joint Chiefs of Staff.

Now most people have never heard of the JROC. There are likely some who wish they never had. The mission of this important panel seems simple—to "provide advice and assessment on military capability needs"—but the process is anything but benign. It requires guiding thousands of requirements, procedures, and people. That can be difficult on any day. But in the middle of two wars where assets are needed yesterday, Ed did an outstanding job of getting vital equipment to those on the frontlines quickly.

Of course, we can't overlook the other half of the Giambastiani team. Cindy, after growing up the daughter of a career Air Force officer, and after 31 years of marriage and many moves with Ed, I suspect that you're ready to settle down in one place for awhile. For those of you who may not know, Cindy has received many awards and honors for her public service. She is the ship's sponsor of the USS New Mexico, and has been very active in the submarine community. She even wrote a cookbook to commemorate the 100th anniversary of the submarine force. I understand that proceeds from the sales go to a scholarship for the children of submariners. Thank you, Cindy, for a lifetime of service and support to your family and to our nation.

Ed, Cindy, thank you for always putting your country, and those around you before yourself. We have been honored to serve with you.

# September 11th Wreath-Laying Ceremony

*Pentagon, 9/11 Cornerstone, Tuesday, September 11, 2007*

THANK you, General Pace.

Scripture teaches us that "those who walk uprightly enter into peace; they find rest as they lie in death." "The souls of the just are in the hand of God, and no torment shall touch them."

On this day, in this place, we gather to offer a tribute to those whose lives were lost six years ago. Friends and family members, this is a day for quiet contemplation, an occasion for you to remember all the times—both good and bad—that you shared with your loved one. We have rebuilt this building, just as many of you have rebuilt your lives and learned to live with your loss. And while time has the ability to heal wounds, to soothe anguished spirits, it can never fully dull the pain or eliminate the awful memories that will forever be associated with this day.

Today the entire nation joins with you. You have never been, and never will be, alone in your sorrow. Those who did not return home that day were more than just friends and colleagues, brothers and sisters, sons and daughters, mothers and fathers. They were fellow Americans, members of our nation's family. They will always be honored as such.

Here at the Department of Defense, we pay an ongoing tribute with our firm commitment to defend the United States against any and all enemies—wherever they may exist.

And let there be no doubt that anyone wishing to revisit harm upon this country will find, in the men and women of this Department, adversaries who have found a clarity of purpose in their grief, a strength of resolve in their anger. The enemies of America—the enemies of our values and our liberty—will never again rest easy, for we will hunt them down relentlessly and without reservation.

We know from history that there will always be those who are impervious to reason or accommodation—enemies who will stop at nothing to do us harm. We, too, will stop at nothing to defend this nation, its citizens, and our values.

As we contemplate the challenges ahead, together we share a profound sense of duty, brought once again into sharp relief today. It is a duty to our nation. A duty to our fellow countrymen. And it is a duty to those who perished here six years ago.

## World Forum on the Future of Democracy

*Williamsburg, VA, Monday, September 17, 2007*

THANK you, Senator Warner.

Senator Warner is a special friend. He has introduced me to the United States Senate for confirmation four times. The first time was more than 20 years ago. And that dates us both. He is a great Virginian, a great American, and we will certainly all miss him when he brings his remarkable career in public service to a close next year.

I want to thank Justice Sandra Day O'Connor, one of the most distinguished jurists and public servants in America, for inviting me today. It was Justice O'Connor who administered my oath of office as Director of Central Intelligence in 1991. And last year we served together on the Baker-Hamilton Commission. Little did I know that my sojourn to Iraq a little over a year ago with the group would be only the first of many such visits for me.

Justice O'Connor and I share something else in common—a love of the College of William and Mary, where she is currently the chancellor. And of course, it was a special pleasure to see her four months ago when I had the honor of giving the commencement address at my alma mater. Attending college in here Williamsburg shaped my love of history and my belief that public service is a vital component of a working democracy—and of a meaningful life.

This setting is fitting for my topic today: a "realist's" view of promoting democracy abroad.

I had quite a reputation as a pessimist when I was in the intelligence business. A journalist once described me as the Eeyore of national security—able to find the darkest cloud in any silver lining. I used to joke that when an intelligence officer smelled the flowers, he'd look around for the coffin. Today, as one looks around the world—wars in Iraq and Afghanistan, an ambitious and fanatical theocracy in Iran, a nuclear North Korea, terrorism, and more—there would seem to be ample grounds to be gloomy.

But there is a different perspective if we step back and look at the world through a wider lens—a perspective that shows a dramatic growth in human freedom and democracy in just the time since this fall's college freshmen were born. Since 1989, hundreds of millions of people—from Eastern and Central Europe and the former Soviet Union, to South Africa, Afghanistan, Iraq, and elsewhere—have been liberated: they have left the darkness of despotism and walked into the bright sunshine of freedom.

Many have seized the opportunity, and freedom has prospered and strengthened; others liberated from the yoke of tyrannical ideologies or dictators continue to struggle to fully realize the dream. At no time in history, though, has freedom come to so many in so short a time. And in every case, the United States, overtly or covertly, in large ways or small, played a role in their liberation.

Still, we Americans continue to wrestle with the appropriate role this country should play in advancing freedom and democracy in the world. It was a source of friction through the entire Cold War. In truth, it has been a persistent question for this country throughout our history: How should we incorporate America's democratic ideals and aspirations into our relations with the rest of the world? And in particular, when to, and whether to try to change the way other nations govern themselves? Should America's mission be to make the world "safe for democracy," as Woodrow Wilson said, or, in the words of John Quincy Adams, should America be "the well-wisher of freedom and independence of all" but the "champion and vindicator only of our own?"

During my time today, I'd like to put this question and its associated debates in some historical context—a context I hope might help inform the difficult policy choices our nation faces today.

Let me first speak to geography—this place we are in.

It is a strange quirk of history that a backwoods outpost in an unexplored corner of America would hold in it the seeds of a global movement toward liberty and self-governance—toward the democratic institutions that underpin the free nations of the world and give hope to countless people in many others.

So much of what defines America first took root here in Virginia along the banks of the James River. When you think about it, the initial impetus for these institutions owed as much to the struggle for survival as to anything else. The challenges were myriad: along with disease, hunger, and war, the settlers faced no small number of divisions and discord. Four hundred years removed from those early days, it is all too easy to forget about these stormy beginnings.

The revolution that brought about this nation was similarly chaotic. As my distinguished William and Mary classmate, the historian Joe Ellis, wrote in his book, Founding Brothers, "No one present at the start knew how it would turn out in the end. What in retrospect has the look of a foreordained unfolding of God's will was in reality an improvisational affair in which sheer chance, pure luck—both good and bad—and specific decisions made in the crucible of specific military and political crises determined the outcome." Ellis further wrote "the real drama of the American Revolution ... was its inherent messiness. This ... exciting but terrifying sense that all the major players had at the time—namely, that they were making it up as they went along, improvising on the edge of catastrophe." We would do well to be mindful of the turbulence of our own early history as we contemplate the challenges facing contemporary fledgling democracies struggling to find their footing.

When I retired from government in 1993, it seemed that the success and spread of democracy was inexorable, a foregone conclusion—that with the collapse of the Soviet Union, the evolution of political systems had reached, in the words of one scholar at the time, the "end of history." But the relative calm in the immediate aftermath of the Cold War served only to mask new threats to the security of democratic nations: ethnic conflicts, new genocides, the proliferation of weapons of mass destruction—especially by rogue states and, above all, a new, more formidable, and more malignant form of terrorism embraced by Islamic extremists.

These new threats, and in particular, the conflicts in Iraq and Afghanistan, and the wider challenge of dealing with radical jihadist movements since September 11th, once again have people talking about the competing impulses in U.S. foreign policy: realism versus idealism, freedom versus security, values versus interests.

This is not a new debate. Not long after winning our own independence, the U.S. was faced with how to respond to the French Revolution—an issue that consumed the politics of the country during the 1790s. The issue was whether to support the revolutionary government and its war against an alliance of European monarchies led by Great Britain. To many, like Thomas Jefferson, the French Revolution, with its stated ideals of liberty, equality, and fraternity, seemed a natural successor to our own. Jefferson wrote that "this ball of liberty, I believe most piously, is now so well in motion that it will roll round the globe."

John Adams and the Federalists, however, were just as adamantly opposed. They were appalled by the revolution's excesses and feared the spread of violent French radicalism to our shores. In fact, they accused the Jeffersonians of being "pimps of France," who "represented cutthroats who walk in rags." The Federalists mocked Jefferson for his rhetorical defense of freedom and equality across the Atlantic while he continued to own slaves. Adams and Alexander Hamilton were, in turn, accused of being crypto-monarchists.

It was left to President George Washington to resolve the matter. He had said that: "My best wishes are irresistibly excited whensoever, in any country,

I see an oppressed nation unfurl the banners of freedom." But the European wars and, in particular, our estrangement from the British, had begun to disrupt the lives of ordinary Americans by impeding trade and causing riots and refugees. Washington, understanding the fragility of America's position at the time, adopted a neutrality policy toward France and would go on to make a peace treaty with Great Britain—sparking massive protests and accusations of selling out the spirit of 1776.

Consider the great historic irony: The United States had recently broken free of the British monarchy only with the help of an absolutist French king. Yet when France itself turned in the direction of popular rule and was confronted by Europe's monarchies, the United States took a pass and made amends with our old British foe.

In short, from our earliest days, America's leaders have struggled with "realistic" versus "idealistic" approaches to the international challenges facing us. The most successful leaders, starting with Washington, have steadfastly encouraged the spread of liberty, democracy, and human rights. At the same time, however, they have fashioned policies blending different approaches with different emphases in different places and different times.

Over the last century, we have allied with tyrants to defeat other tyrants. We have sustained diplomatic relations with governments even as we supported those attempting their overthrow.

We have at times made human rights the centerpiece of our national strategy even as we did business with some of the worst violators of human rights. We have worked with authoritarian governments to advance our own security interests even while urging them to reform.

We have used our military to eliminate governments seen as a threat to our national security, to undo aggression, to end ethnic slaughter, and to prevent chaos. In recent times, we have done this in Grenada, Panama, Kuwait, the Balkans, Haiti, Afghanistan, and Iraq. In the process, we have brought the possibility of democracy and freedom to tens of millions more who had been oppressed or were suffering.

To win and protect our own freedom, the United States has made common cause with countries that were far from free—from Louis XVI, to one of history's true monsters, Joseph Stalin. Without the one there is no American independence. Without the other, no end to the Third Reich. It is neither hypocrisy nor cynicism to believe fervently in freedom while adopting different approaches to advancing freedom at different times along the way—including temporarily making common cause with despots to defeat greater or more urgent threats to our freedom or interests.

The consuming goal of most of my professional life was containing the threat of the Soviet Union and seeing a Europe made whole and free. For most of the Cold War, the ideal surely seemed distant, even unreachable.

One prominent columnist wrote in Time magazine in 1982 that "It would be wishful thinking to predict that international Communism someday will either self-destruct or so exhaust itself."

During that struggle, as for most of our history, inspiring presidential rhetoric about freedom, along with many firm stands for human rights and self-determination, had to coexist with often grubby compromises and marriages of convenience that were necessary to stave off the Evil Empire.

But the Western democracies—joined as the Atlantic Alliance—came together to get the big things right. The democracies' shared belief in political and economic freedom and religious tolerance was the glue that held us fast despite the many quarrels along the way.

President Bush said in his second inaugural address, "[I]t is the policy of the United States to seek and support the growth of democratic movements and institutions in every nation and culture, with the ultimate goal of ending tyranny in our world."

When we discuss openly our desire for democratic values to take hold across the globe, we are describing a world that may be many years or decades off. Though achievement of the ideal may be limited by time, space, resources, or human nature, we must not allow ourselves to discard or disparage the ideal itself. It is vital that we speak out about what we believe and let the world know where we stand. It is vital that we give hope and aid to those who seek freedom.

I still remember working on the advance team for President Ford when he attended the Helsinki conference in 1975. Many critics were opposed to America's participation, since they believed that the accords did little but ratify the Soviet Union's takings in Eastern and Central Europe. The treaty's provisions on human rights were disparaged as little more than window dressing. However, the conference and treaty represent another of history's ironies. The Soviets demanded the conference for decades, finally got it, and it helped destroy them from the inside. We "realists" opposed holding the conference for decades, and attended grudgingly. We were wrong. For the meeting played a key role in our winning the Cold War.

Why? Because the human-rights provisions of the treaty made a moral statement whose significance was not lost on the dissidents behind the Iron Curtain. Helsinki became a spur to action, a rallying cry to fight tyranny from within and plant democracy in its place.

Vaclav Havel later said that the accords were a "shield, a chance to resist coercion and make it more difficult for the forces of coercion to retaliate." Lech Walesa called it a turning point "on the road to change in Gdansk."

President Carter's promotion of the spirit of Helsinki—his elevation of human rights—for the first time in the Cold War denied the Soviet Union the respect and the legitimacy it craved. Ronald Reagan's muscular words—

labeling the U.S.S.R the "Evil Empire" and demanding that Mr. Gorbachev tear down that wall—combined with his muscular defense policies hastened the implosion of the Soviet system.

Did these policies reflect hard-edged realism or lofty idealism? Both, actually. Were they implemented to defend our interests or to spread our democratic values? Again, both.

An underlying theme of American history is that we are compelled to defend our security and our interests in ways that, in the long run, lead to the spread of democratic values and institutions.

Since September 11th, these questions, contradictions, and dilemmas have taken on new urgency and presented new challenges for decision-makers, especially in an information age where every flaw and inconsistency—in words or deeds—is highlighted, magnified, and disseminated around the globe.

And, as with the Cold War, every action we take sends a signal about the depth of our strength and resolve. For our friends and allies, as well as for our enemies and potential adversaries, our commitment to democratic values must be matched by actions.

Consider Afghanistan. The democracies of the West and our partners are united in the desire to see stability and decent governance take hold in a land that was not only Al Qaeda's base of operations, but also home to one of the most oppressive governments in the world. And yet, though there is little doubt about the justness, necessity, and legitimacy of the Afghanistan mission, even though we agree that democracy is key to enduring stability there, many Allies are reluctant to provide the necessary resources and put their men and women in the line of fire.

Afghanistan is, in a very real sense, a litmus test of whether an alliance of advanced democracies can still make sacrifices and meet commitments to advance democracy. It would be a mark of shame on all of us if an alliance built on the foundation of democratic values were to falter at the very moment that it tries to lay that foundation for democracy elsewhere—especially in a mission that is crucial to our own security.

Likewise, for America to leave Iraq and the Middle East in chaos would betray and demoralize our allies there and in the region, while emboldening our most dangerous adversaries. To abandon an Iraq where just two years ago 12 million people quite literally risked their lives to vote for a constitutional democracy would be an offense to our interests as well as our values, a setback for the cause of freedom as well as the goal of stability.

Americans have never been a patient people. Today, we look at Russia, China, Afghanistan, Iraq, and others—and wonder at their excruciatingly slow progress toward democratic institutions and the rule of law.

The eminent French historian Helene Carrere d'Encausse wrote in 1992: "Reforms, when they go against the political traditions of the centuries, cannot be imposed in a hurry merely by enshrining them in the law. It takes time, and generally they are accompanied by violence." She added: "Reforms that challenge centuries of social relations based on ... the exclusion of the majority of society from the political process, are too profound to be readily accepted by those who have to pay the price of reform, even if they are seen to be indispensible. Reforms need time to develop ... It is this time that reformers have often lacked."

For more than 60 years, from Germany and Japan to South Korea, the Balkans, Haiti, Afghanistan, and Iraq, we and our allies have provided reformers—those who seek a free and democratic society—with time for their efforts to take hold. We must be realists and recognize that the institutions that underpin an enduring free society can only take root over time.

It is our country's tragedy, and our glory, that the tender shoots of freedom around the world for so many decades have been so often nourished with American blood. The spread of liberty both manifests our ideals and protects our interests—in making the world "safe for democracy," we are also the "champion and vindicator" of our own. In reality, Wilson and Adams must coexist.

Throughout more than two centuries, the United States has made its share of mistakes. From time to time, we have strayed from our ideals and have been arrogant in dealing with others. Yet, what has brought us together with our democratic allies is a shared belief that the future of democracy and its spread is worth our enduring labors and sacrifices—reflecting both our interests and our ideals.

I would like to close by returning to this corner of Virginia. In September 1796, shortly before George Washington left office, he addressed in his farewell statement an American people who had passed through the dangerous fires of war and revolution to form a union that was far from "perfect," but was a historic accomplishment nonetheless. He told them: "You have, in a common cause, fought and triumphed together; the independence and liberty you possess are the work of joint councils and joint efforts, of common dangers, sufferings, and successes."

In this historic place, among old friends and new, let us take time to reflect on the common causes in which we have fought and triumphed together—to protect our own liberty, and to extend its blessings to others. As we prepare for the challenges ahead, let us never forget that together we will face common dangers, sufferings, and successes—but with confidence that, together, we will continue to protect that tender shoot of liberty first planted in this place so long ago.

Thank you.

# United States Air Force 60th Anniversary

*Pentagon Courtyard, Tuesday, September 18, 2007*

GENERAL Moseley, Secretary Wynne, thank you for inviting me to this celebration. As Second Lieutenant Gates at Whiteman Air Force Base 40 years ago, I would never have imagined being on the same stage with the Air Force Chief of Staff and the Secretary of the Air Force. So it is a real honor. When I was commissioned at Lackland in January of 1967, we were all asked if we would have any guests at the ceremony in the rank of Colonel or GS-15 or above who would warrant VIP treatment. None of us had friends in such high places.

Colonel Day, Congressman Spratt, Congresswoman Wilson, Congressman Stearns, other distinguished guests. Members of the Department. Airmen, past and present.

Ever since the dawn of civilization, the idea of flight has held an unshakeable grip on the human imagination. The myths of ancient Greece, the musings of great philosophers, the charcoal sketches of Leonardo da Vinci—all illustrated the dream that one day mankind would travel in the skies and maybe even among the stars.

For the romantics, it was always a vision of the future. For the cynics, it was little more than the fantasy of an overactive mind.

Nearly four millennia of dreams and fantasies finally gave way to reality in an unlikely place, the result of years of tireless work by an unlikely pair: two brothers from Ohio, bike salesmen by day, aeronautic visionaries by night.

That first tentative and halting foray into the sky by a heavier-than-air flying machine—a mere ten feet above the sands of Kitty Hawk, 120 feet across the dunes—marked more than just the dawn of the Age of Flight. It also marked the beginning of the incredible journey that brings us here today in celebration of the 60th anniversary of the United States Air Force.

It was not an easy journey. The infancy of flight was beset with problems. After a failed attempt at flight in 1901, Orville Wright recalled his brother saying that man would not fly for a thousand years. And shortly thereafter, a famous scientist said that "Flight by machines heavier than air is unpractical and insignificant, if not utterly impossible." As in previous ages, however, there were those who firmly believed that aviation could change the face of the world. It fell to a man named Billy Mitchell to fight the prevailing conventional wisdom within the military establishment. He did so with great fervor, and little tact. Senior officers took to calling him the "Kookaburra," an Australian bird more commonly known as the "laughing jackass."

Mitchell was eventually court-martialed, and one of his protégés took over the cause within the military. For his determination, that young man, who had learned to fly with the Wright Brothers, was finally given the choice of resigning from the services or being court-martialed. He chose the court-martial, but was instead sent into exile. He eventually returned to good favor and ended up making something of a name for himself. Some of you may have heard of five-star General "Hap" Arnold.

Despite the early arguments and acrimony, despite myriad questions over airpower's utility in war, the public quickly embraced America's intrepid flying men. Millions came to know the great aces of two world wars—Eddie Rickenbacker, Dick Bong, and others. And they came to know entire units like the Flying Tigers and the Doolittle Raiders—units that sustained and inspired a nation during some of its darkest and most trying days.

Since then and throughout the 60-year history of the Air Force, the American people have stood in awe as Airmen continued to push the limits of bravery and endurance—as they crashed through the sound barrier many times over, and extended the range, scope, and nature of air missions beyond what anyone could have imagined.

And the Air Force has continued to inspire us—during the Berlin Airlift in 1948, during two wars in Asia, during the long twilight struggle against the Soviet Union, and today in a new conflict that tests our ingenuity and creativity as much as, if not more than, our technological prowess.

I seriously doubt anyone would have believed that, in the 21st century, Airmen on horseback in the wilds of Afghanistan would direct B-52s to provide close air support for cavalry charges—ironic, considering that early doubters of aviation's military value feared that noise from the planes would frighten the cavalry's horses. And then there is the nontraditional nature of the Air Force's mission in Iraq, where Airmen and airpower are used as often to protect as to destroy. Curtis LeMay is no doubt spinning in his grave.

Above all else, it is the men and women of the Air Force who have for so many years made this institution what it is—the sword and shield of the nation, its sentry and its avenger.

I want to extend my personal thanks to the distinguished guests recognized earlier. For these men and women, the allure of the sky proved stronger than the forces—of both man and nature—working to keep them on the ground. Many had to defy personal fears. Others had to defy societal prejudice. All demonstrated unflagging devotion. They are examples for us all.

As we look back on everything that has made this birthday possible, let us also look forward to many more birthdays as the Air Force continues its dominance of air, space, and cyberspace.

You represent the best that our military and our nation has to offer—a long and distinguished heritage of courage, and endless horizons of innovation.

Thank you.

## POW/MIA Recognition Day

*Pentagon Parade Field, Friday, September 21, 2007*

Good morning.

Thank you all for coming—veterans and MIA/POW groups, leaders of this department, and distinguished guests.

I want to extend a special thanks to two groups of people. To the former prisoners of war—your presence here is an important gesture of solidarity, tangible proof that we as a nation will continue to honor all of those who have been captured or gone missing. You have been through a great ordeal and may still carry scars. Know always that our nation is keenly aware of, and ever thankful for, your sacrifice.

I would also like to thank the families and friends of those still missing. Missing-in-action status is marked by ambiguity and uncertainty—a severe test of spirit and resolve for anyone seeking closure. Your attendance today proves once again that the bond of love transcends the passage of time—that while our nation's heroes may remain missing in body, they are always present in spirit.

We will neither forget our duty to bring home all POWs and MIAs, nor relent in our efforts to do so.

Last week, our nation reflected on the sixth anniversary of the September 11th attacks. It was, for most of us at the Pentagon, a day of both remembrance and recommitment. Even as we mourned those who were killed, we also recalled the clarity of purpose and strength of resolve we found in the aftermath of the attacks.

I mention this because today, also, is a day of remembrance and recommitment. And it is furthermore a day that will always be connected to September 11th by coincidence of the calendar. When mid-September was first chosen to honor and remember our POWs and MIAs, it was picked because it was unconnected to any specific war or cause. It was meant to be a day wholly its own.

I believe, however, that it is in some ways fitting that this day should come just after our annual September 11th remembrance. Throughout our nation's history, it has always fallen to the men and women of the Armed Forces to respond to aggressors and adversaries. To endure arduous and

Spartan conditions; to risk life and limb on the battlefield; to make the sacrifices that are, in the final analysis, both our nation's tragedy and our glory.

As in past eras, we have once again been called to duty in a conflict that is global in scope, and generational in duration. And, as in the past, the honor, courage, and sacrifice of our men and women in uniform will be our nation's glory. As in the past losing them on the battlefield is ever our tragedy.

The enemies we face today—and the ideology that inspires them—are in some ways similar to earlier foes. Their ambitions are global, their hopes totalitarian. But they are also very different. They have no borders to defend, diplomats to negotiate with, or armies, navies, or air forces to defeat. Their instrument is terror; their victims, the innocent. And they don't take prisoners.

Even with all the advances and advantages of technology, the vagaries of war still mean that we cannot account for everyone—even in the 21st century.

Today, I would like to pay special tribute to four soldiers who have gone missing in Operation Iraqi Freedom: Staff Sergeant "Matt" Maupin; Specialist Ahmed al-Taei; Specialist Alex Jimenez; Private Byron Fouty. They may not be well known to the public, but within the brotherhood of arms, they will never be forgotten, or left behind.

These men are the latest additions to the ranks of tens of thousands who remain missing from previous conflicts. And they are the latest additions to the ranks of those we remember today.

In a few minutes, we will hear from Patricia Scharf, an icon here at the Pentagon, where she has worked for 37 years. Mrs. Scharf's husband, Charles, was shot down in Vietnam in 1965, and last year his remains were identified and buried less than half a mile from here, in Section 66 of Arlington National Cemetery. Her words will undoubtedly capture the feelings many of you have far better than my own.

Before she speaks, it is my distinct honor to introduce the Chairman of the Joint Chiefs of Staff, Marine General Peter Pace.

General Pace's distinguished career is too long to go into here, but I would like to mention one thing. Many of you may have heard that General Pace keeps on his desk a photograph of the first man lost under his command. It is not as well-known that he also keeps a picture of the first serviceman to go missing in the War on Terror. For General Pace, those pictures are a reminder not only of the terrible cost of war, but also of his solemn duty to all our troops in harm's way.

In General Pace, our men and women in uniform have a tireless ally and advocate. And in General Pace, our nation has been blessed to have the services of a dedicated and decorated Marine. When he retires at the end of this month, I will sorely miss his counsel, his candor, and his friendship.

General Pace.

## Senate Appropriations Committee

*106 Dirksen Senate Office Building, Washington, DC, Wednesday, September 26, 2007*

Mr. Chairman, members of the Committee:

First, I'd like to thank the Committee for all you have done to support our military over these many years. And I appreciate the opportunity to speak with you today about the Fiscal Year 2008 Global War on Terror Request.

Mr. Chairman, thank you for your kind words about General Pace. I've come to trust him completely, and relied on his advice these past 10 months. And I thank you for joining me in wishing him well and thanking him for his four decades of service to our country.

I urge the Congress to approve the complete Global War on Terror Request as quickly as possible and without excessive and counterproductive restrictions. That will help the Department manage its expenses.

While this hearing is focused on the war funding request, I would like to note, with concern, the Committee's recent report language of the Defense Appropriations Bill, concerning Section 1206, Global Train and Equip. This authority is a unique tool that provides commanders a means to fill the long-standing gaps in our ability to build the capacity and capabilities of partner nations. It has become a model of interagency cooperation between the State and Defense Departments, both in the field and here in Washington. Secretary Rice and I both fully support this authority. Its benefits will accrue to our successors in future administrations, and I urge the Committee to reinstate our full request for $500 million in the base budget and continue support in future years.

I would also like to voice my strong support today to the Defense—to the State Department's portion of the War on Terror request. As you know, the challenges we face in Iraq, Afghanistan and elsewhere are fundamentally political, economic and cultural in nature, and are not going to be overcome by military means alone. It will be very difficult for our troops and their commanders to succeed without the key non-military programs and initiatives included in the request for the State Department.

The initial FY 2008, Fiscal Year 2008 War on Terror funding request for the Department of Defense was submitted in February for $141.7 billion. At the time, the Department stated that this initial request was an estimate based on a straight-line projection of ongoing war costs and would need to be adjusted given the evolving and dynamic realities on the ground in both Iraq and Afghanistan.

Major elements of that initial request included:

- $70.6 billion for operations, including incremental pay, supplies, transportation, maintenance and logistical support to conduct military operations.
- $37.6 billion to repair and replace equipment that has been destroyed, damaged or stressed by the ongoing conflicts.
- $15.2 billion for force protection, including new technologies and equipment to protect troops from Improvised Explosive Devices and other threats.
- $4.7 billion to train and equip Afghan and Iraqi security forces.
- $1 billion dollars for the Commander's Emergency Response Program, funds that can be dispensed quickly and applied directly by U.S. commanders for local needs.

The department submitted its first adjustment on July 31st, 2007, for $5.3 billion to buy 1,520 Mine Resistant Ambush Protected, MRAP, vehicles, bringing the total War on Terror request to $147 billion.

The second adjustment, to be submitted by the president, seeks approximately $42 billion, bringing the total FY 08 DoD request to nearly $190 billion. The second adjustment includes:

- $6 billion to support the Army and Marine combat formations currently in Iraq through fiscal year 2008—this takes into account the President's announced intention to redeploy five Army Brigade Combat Teams by next summer.
- $14 billion for force protection, $11 billion of which will go toward fielding approximately 7,000 more MRAP vehicles on top of the 8,000 already funded or requested—this also includes funding to better defeat enemy snipers and to modify Army combat vehicles to improve survivability.
- $9 billion for reconstitution to ensure that we provide our forces the critical equipment and technology they need for future combat operations.
- $6 billion for training and equipment that will accelerate the deployment readiness of Army units—this includes a billion dollars to support the National Guard pre-deployment training.
- $1 billion dollars to improve U.S. facilities in the region and consolidate our bases in Iraq.
- $1 billion dollars to train and equip Iraqi Security Forces.

Mr. Chairman, I know that Iraq and other difficult choices America faces in this war on terror will continue to be a source of friction within the Congress, between the Congress and the President and the wider public debate. Considering this, I would like to close with a word about something I believe we can all agree on—the honor, courage and great sense of duty we have witnessed in our troops.

Under some of the most trying conditions, they have done far more than what was asked of them and far more than what was expected. Like all of you, I am both humbled and inspired by my trips to Walter Reed, Bethesda and other military hospitals, and to the frontlines in Iraq and Afghanistan.

And like all of you, I always keep our troops, their safety and their mission, foremost in my mind every day. Once again, I thank each of you and the rest of the Congress for the support you have given them and their families during this period of great consequence for America.

Thank you, Mr. Chairman.

## Farewell Ceremony for General Peter Pace

*Summerall Field, Fort Myer, VA, Monday, October 01, 2007*

Mr. President, Mr. Vice President, distinguished guests. Admiral Mike Mullen—thank you for your willingness, yet again, to answer your country's call. Men and women of the Armed Forces.

Even a condensed version of General Pete Pace's 40 years of distinguished service would take up much of the morning. So I would like to spend a few minutes talking about the man I have come to know over the past ten months. I should begin with a story about a place and a platoon.

At 3:40 A.M., on January 31st, 1968, thousands of North Vietnamese soldiers engulfed the ancient city of Hue as part of what became known as the Tet offensive—leaving a small contingent of American and South Vietnamese surprised and surrounded.

Several miles down the road at Phu Bai, members of Golf Company, 2nd Battalion, 5th Marines, were called in to the rescue—for what one officer called "an afternoon of street fighting."

That "afternoon" turned into a month long campaign in a city where, according to one officer, "every window, every roof, and every intersection harbored potential death."

By the end of the first week, the American flag flew proudly above the enemy headquarters on the south side of the city.

But for 2nd Platoon, the cost of this real estate had been high: Two-thirds of their comrades could no longer bear arms, including their leader. Many fell in the opening minutes of the battle while trying to take a bridge. And there was much more fighting left.

Such was the scene that awaited Golf Second of the Fifth's newest platoon leader, the third in as many weeks: Second Lieutenant Peter Pace, age 22. Such was the heroism and the hope of those who would begin to shape that young Marine into the officer we honor today.

The first thing Pete did was call together the squad leaders and say to them, "My name is Pete Pace, and I have no idea what I'm doing... If you guys will help me out and talk to me, I promise that I will listen."

Barney Barnes was one of Pete's squad leaders. He recalled, "Some officers come in and they demand respect. General Pace didn't do that. He earned our respect. He earned it by loving us, caring for us, teaching us, making sure that we were the best Marines that we could possibly be." Many years later, of those men, and of that experience, Pete said, "It was their blood that gave me a debt that I can never fully repay."

Marine officers take special care to consider themselves leaders of Marines first, and whatever military specialty they might be second. And General Pace is certainly that. But as we have seen through 40 years of extraordinary service, this proud Marine has also shown himself to be a gifted leader of soldiers, sailors, airmen—and with his wife Lynne, a firm advocate for their families as well.

General Pace brings his career to a close as one of the last of a dwindling breed of officers. His four decades in uniform have spanned four eras of the U.S. military's modern history. From:

- The Vietnam War and the draft;
- To the all-volunteer force and victory in Desert Storm;
- To the false tranquility following the Cold War; and
- Then the post-9/11 campaigns our Armed Forces have waged in Afghanistan, Iraq, and against violent jihadists worldwide.

I should note that about the same time Pete was rallying his men in Vietnam, I was just beginning my transition from the Air Force to the CIA. While Pete was getting shot at, I was starting my CIA training here, having just transferred from duty tending ICBMs at Whiteman Air Force Base. The closest I've been to live combat is going to the Hill to testify, which is why I've always wanted Pete there by my side.

When I arrived at the Pentagon, Pete helped ensure that I had everything I needed to lead this department.

His expertise and wisdom across a wide variety of complex issues has helped guide my every decision. I value his candor, and I have come to trust his judgment in all matters. His sense of humor has lifted me when I was down and his sense of duty has driven me to do right by the men and women of this Department. He has been more than a mentor. He has been a friend.

And the thing that sticks with me is that although General Pace is the Chairman of the Joint Chiefs of Staff, the most powerful military officer in the world, he still cares for everyone in our Armed Forces as if he were still their platoon leader. Whenever we at the highest levels are faced with a decision, no matter how big or small, he will always ask aloud the question that has guided him throughout his career, "How will this impact Private First Class Pace and Mrs. Pace?" In General Pace, the men and women of the Armed Forces have had a leader who never lost sight of the individual or the troops on the frontlines.

Through it all, he has carried himself with humility, dignity, and grace—qualities that were on display when he joined those battle-weary Marines in the rubble of an ancient city halfway around the world and incurred, as he said, a debt he could never fully repay.

Pete, I speak for everyone in the nation when I say that your debt has been more than repaid. In my service under seven presidents, I have been privileged to serve with great leaders. You are one of the finest.

You leave today as you should: With flags flying, words of tribute ringing in your ears, and the heartfelt thanks of a grateful nation.

We wish you and Lynne the very best. Thank you.

## Association of the United States Army

*Washington, DC, Wednesday, October 10, 2007*

First, I would say welcome to Washington. A city where those who travel the high road of humility encounter littler traffic. Where people often say, "I'll double-cross that bridge when I come to it." Where you can see prominent people walking down lover's lane holding their own hands.

The story Peter told about my wanting to be a doctor is true. I often tell people my decision to join CIA probably saved countless lives.

I returned Saturday night from visiting five Latin American countries, including Peru. My visit there reminded me of the perils facing hosts when receiving foreign dignitaries. Some years ago, a European foreign minister, a notoriously heavy drinker, was visiting Peru. He was at a formal event, and he was drunk. Music was playing, someone in a long, flowing gown passed him. The foreign minister asked this person to dance. The individual turned on the drunken minister and somewhat haughtily replied, "First, you are drunk. Second, this is not a waltz, this is the Peruvian national anthem. Third, I am not a woman but the Cardinal Archbishop of Lima."

I leave tonight for Russia, where all visitors for decades have assumed they were being spied on. Often, visitors have been a little too paranoid. Such as the time Canadian hockey player Phil Esposito was in Moscow and he and his roommate decided to find the bug in their hotel room. They searched high and low to no avail. And they flipped the rug back and found there the supposed bug inset in the floor. With great effort, they unscrewed it only to hear a thunderous crash. They had undone the anchor of the chandelier in the room below them.

Despite the travel, different time zones and different so on, perhaps the most difficult, confusing aspect of this job for me is closer to home. It may also have vexed some of you at some point and that is navigating the Pentagon—in every sense of the word. General Eisenhower learned this the hard way shortly after World War II when he tried to return to his office—by himself—after eating at the general officers' mess. Eisenhower later wrote: "So hands in pockets and trying to look as if I were out for a carefree stroll around the building, I walked … and walked and walked, encountering neither landmarks nor people who looked familiar. One had to give the building grudging admiration; it had apparently been designed to confuse any enemy who might infiltrate it."

Newsman David Brinkley used to tell a story of the early days at the Pentagon. A woman told a Pentagon guard she was in labor and needed help in quickly getting to a hospital. The guard said, "Madam, you should not have come in here in that condition." She answered, "I wasn't in this condition when I came in."

I'm honored to be invited to this important forum, and to be with an organization that has done so much to support soldiers and their families. In fact, AUSA was one of the very first speaking events I accepted back in January.

I should start by saying that even before assuming my current job, the state of the Army was one of my chief concerns. Since then, I've had to sign orders extending deployments and sign letters of condolences to the families of the fallen. And then there are the visits with the wounded at Walter Reed, Bethesda, Brooke, Tripler, and at Landstuhl. To be honest, before I went to Walter Reed the first time, I dreaded it, not knowing how I or the wounded would react. But people told me, "No, they'll lift you up." And they were so right.

Through it all I have never forgotten that we are talking about individuals—America's sons and daughters—and not numbers on a press release or a web page. To me it's not institutional, it's very personal.

Just a few days ago we swore in a new Chairman of the Joint Chiefs, Admiral Mike Mullen. One of the things that convinced me that Admiral Mullen was the right man for the job was when he was asked, as Chief of Naval Operations, what his top concern was. He didn't start talking about a new air craft carrier or submarine. He said: "The Army."

That says a good deal about Admiral Mullen and the priorities of the leadership of the Department of Defense

These past couple of days you've heard from much of the Army senior leaders about their plans and goals for the service. Today, I'd like to offer my perspective on where the Army stands, and where it needs to be headed as it resets from the current conflicts and reshapes itself for the future.

I would like to frame the discussion in two ways:

- First, what America owes the Army after six years of war; our first protracted conflict with an all-volunteer force since the American Revolution.

- And second, what the Army owes America—as it prepares to defend this country's freedom and interests in the decades ahead.

The U.S. Army today is a battle-hardened force whose volunteer soldiers have performed with courage, resourcefulness, and resilience in the most grueling conditions. They've done so under the unforgiving glare of the 24 hour news cycle that leaves little room for error, serving in an organization largely organized, trained, and equipped in a different era for a different kind of conflict. And they've done all this with a country, a government—and in some cases a defense department—that has not been placed on a war footing.

As a result of this stress, there has been a good deal of concern about the condition of the Army, leading some to speculate that it is "broken." I think not.

On numerous occasions, skeptical reporters have come back from Iraq and Afghanistan amazed at the high morale and discipline they see in our soldiers. Recruiting has been no small challenge, but targets are being met. The high retention rates continue to be nothing short of remarkable, especially when considering that those most likely to re-enlist are those most often deployed. For all that is given up to be in this line of work, our soldiers gain something that few can claim—they know that they are defending our country and shaping the course of history. That's no small thing, and it is a source of great pride.

But while the Army certainly is not broken, it is under stress, and, as General Casey puts it, "out of balance."

So when one considers what the nation owes the Army, the answer is a good deal. And it starts with gratitude and appreciation for the service and sacrifice of soldiers and their families.

America has come a long way on this front from the late 1960s and early 1970s during our last protracted and controversial war. You see it in airports all over the country, where soldiers are met with standing ovations by passengers in the terminal. I've been there and seen it myself. There are free meals and rounds of drinks. And, above all, simple thank yous. The appreciation is real, it is heartfelt, and it bridges any political divide.

For those who have made the ultimate sacrifice, the country owes their families every care and benefit as they make the wrenching transition to life without a father or mother, brother or sister, daughter or son.

To the wounded we have a moral obligation to see that the superb life-saving care they receive in the theater and at Landstuhl is matched by the outpatient treatment that will allow them to transition smoothly to the next phase in their lives—drawing on support that facilitates that transition, not impedes it. The lapses that have occurred in this area will not be tolerated nor repeated.

We are well familiar with the high pace of deployments and the strain this has placed on soldiers and their families. There are units like the 2nd Brigade of the 10th Mountain Division, now finishing up its 15th month in Iraq. Since 9/11 no other Army brigade has spent more time away from home.

But it is also important for the men and women of the Army to know that relief is on the way:

- While U.S. forces will play some role in Iraq for years to come, a reduction in the size of our commitment there is inevitable. Most of the serious discussion today is over how and when;
- The Army is expanding by some 65,000 soldiers, and I am prepared to support plans to speed up that process as long as we can do it without sacrificing quality;
- With strong bipartisan support in the Congress, tens of billions of dollars have been allocated to reconstitute damaged and destroyed equipment; and
- New programs and resources are coming on line to make the Army's covenant with families a reality.

America's ground forces have borne the brunt of under-funding in the past and the bulk of the costs—both human and material—of the wars of the present. By one count, investment in Army equipment and other essentials was underfunded by more than $50 billion before we invaded Iraq. By another estimate, the Army's share of total defense investments between 1990 and 2005 was about 15 percent. So resources are needed not only to recoup from the losses of war, but to make up for the shortfalls of the past and to invest in the capabilities of the future.

How those resources are used, and where those investments are made today will shape the Army for decades to come. We do not get the dollars or the opportunity to reset very often. So it's vital we get it right.

This will call on accountable and visionary leadership across the service and up and down the chain of command.

One of the Army's concerns you've heard about at this conference is getting back to training for "high intensity" situations—a capability vitally important to deter aggression and shape the behavior of other nations.

It strikes me that one of the principal challenges the Army faces is to regain its traditional edge at fighting conventional wars while retaining what it has learned—and relearned—about unconventional wars—the ones most likely to be fought in the years ahead.

One of my favorite sayings is that "experience is that marvelous thing that enables you to recognize a mistake when you make it again."

In the years following the Vietnam War, the Army relegated unconventional war to the margins of training, doctrine, and budget priorities. Consider that in 1985 the core curriculum for the Army's 10-month Command and General Staff College assigned 30 hours—about four days—for what was is now called low intensity conflict. This was about the same as what the Air Force was teaching at its staff college at the time.

This approach may have seemed validated by ultimate victory in the Cold War and the triumph of Desert Storm. But it left the service unprepared to deal with the operations that followed: Somalia, Haiti, the Balkans, and more recently Afghanistan and Iraq—the consequences and costs of which we are still struggling with today.

The work that has been done to adapt since has been impressive—if not nearly miraculous. Just one example is the transformation of places like the National Training Center, where, as one officer put it, the Army has "cut out a piece of Iraq and dropped it into Southern California," replete with a dozen villages and hundreds of Arab Americans employed as role players. The publication of the counterinsurgency manual is another milestone, and is being validated by the progress we've seen in Iraq over the past few months. This work and these lessons in irregular warfare need to be retained and institutionalized, and should not be allowed to wither on the bureaucratic vine.

Put simply, our enemies and potential adversaries—including nation states—have gone to school on us. They saw what America's technology and firepower did to Saddam's army in 1991 and again in 2003, and they've seen what IEDs are doing to the American military today. It is hard to conceive of any country challenging the United States directly on the ground—at least for some years to come.

Indeed, history shows us that smaller, irregular forces—insurgents, guerrillas, terrorists—have for centuries found ways to harass and frustrate larger, regular armies and sow chaos. As one officer recently told the Washington Post, "the toys and trappings have changed," but the fundamentals have not.

We can expect that asymmetric warfare will remain the mainstay of the contemporary battlefield for some time. These conflicts will be fundamentally political in nature, and require the application of all elements of national power. Success will be less a matter of imposing one's will and more a function of shaping behavior—of friends, adversaries, and most importantly, the people in between.

One of the challenges facing the Army will be how to incorporate the latest in technology without losing sight of the human and cultural dimensions of the irregular battlefield. For example, we have spent billions on tools and tactics to protect against IEDs. Yet, even now, the best way to defeat these weapons—indeed the only way to defeat them over the long run—is to get tips from locals about the networks and the emplacements or, even better, to convince and empower the Iraqis to prevent the terrorists from emplacing them in the first place.

In addition, arguably the most important military component in the War on Terror is not the fighting we do ourselves, but how well we enable and empower our partners to defend and govern their own countries. The standing up and mentoring of indigenous armies and police—once the province of Special Forces—is now a key mission for the military as a whole. How the Army should be organized and prepared for this advisory role remains an open question, and will require innovative and forward thinking.

The same is true for mastering foreign language—a particular interest of mine—and building expertise in foreign areas. And until our government decides to plus up our civilian agencies like the Agency for International Development, Army soldiers can expect to be tasked with reviving public services, rebuilding infrastructure, and promoting good governance. All these so-called "nontraditional" capabilities have moved into the mainstream of military thinking, planning, and strategy—where they must stay.

Finally, there is a generation of junior and mid level officers and NCOs who have been tested in battle like none other in decades. They have seen the complex, grueling face of war in the 21st century up close. They've lost friends and comrades. Some have been deployed multiple times and want to have a semblance of a normal life—get married, start a family, continue their schooling.

These men and women need to be retained, and the best and brightest advanced to the point that they can use their experience to shape the institution to which they have given so much. And this may mean reexamining assignments and promotion policies that in many cases are unchanged since the Cold War.

In closing, I should tell you that when I speak to Army leaders I make it a point to ask them to communicate to their subordinates not only the thanks of a grateful and admiring nation, but also our pride in what they have accomplished.

The story of just one unit explain why.

The 1st Brigade of the First Armored Division, the "Ready First Brigade," had been based in Germany for more than 60 years, most of that time preparing to beat back a Soviet invasion across the Fulda Gap. It was deployed to Iraq in 2003, and extended after the Sadr uprising in 2004.

Last year—before there was a "surge," or a "new way forward," or a new counterinsurgency manual—they were sent back to Iraq, this time to Ramadi. The city was controlled by insurgents and Al Qaeda, and was written off as lost. The brigade commander was told: "fix it, don't destroy it." It was up to him, his staff, and his soldiers to figure out the rest.

And so instead of patrolling from large bases, the Ready First Brigade set up small combat outposts in Al Qaeda strongholds—where troops led by sergeants and lieutenants and captains cleared and held neighborhoods one at a time. The enemy would not go quietly—and responded with an onslaught of roadside bombs, mortars, and ambushes. Among the hundreds of stories of heroism that emerged from this period was of Sergeant David Anderson. He saved the lives of several soldiers on September 24th after they were ambushed and hit by multiple IED attacks. He would later receive the Silver Star for his efforts.

One of the Brigade staff officers was Captain Travis Patriquin. He spoke several languages, including Arabic, and he grew a mustache to fit in. He became the expert on the neighboring tribes—local power brokers going back hundreds of years who had been largely shunned up to that point by our military.

Like any self-respecting army officer, Patriquin had a Powerpoint presentation. It was called "How to Win in Al Anbar by Captain Trav." But instead of charts and graphs, this presentation used stick figures and simple stories to teach soldiers how to deal with Iraqi tribes—a relationship where "shame and honor" meant a good deal more than "hearts and minds." At this young captain's direction, the brigade courted local sheiks over cigarettes and endless cups of tea—outreach that, combined with Al Qaeda's barbarism, helped spark the "Anbar Awakening" that has garnered so much attention and praise in the past months.

Over time, Ramadi was taken back from Al Qaeda and given back to its people. These gains came at no small cost. During its tour, this brigade would suffer more than 95 killed and 600 wounded. One of them was Captain Patriquin. He did not have a chance to see his ideas and efforts bear fruit, but no doubt would have been proud to have seen what the hard work, courage, and ingenuity of the soldiers had started: A city liberated. Al Qaeda uprooted and reeling. And the tide turned, at least in this one important battle, in a conflict that will determine the future of the Middle East for decades to come.

It is soldiers and stories like these—repeated in so many places and so many times—that inspire us and make us proud and hopeful about the future of America's Army. Our country's defense could not be in better hands.

Thank you very much.

## Military Academy of the General Staff

*Moscow, Russia, Saturday, October 13, 2007*

THANK you for that introduction. And thank you for having me here today. My congratulations to General Belousov on his recent appointment to lead this institution.

It is testimony to how far we have come that the Military Academy of the General Staff would invite the former head of the CIA and current American Secretary of Defense to speak in this prestigious setting.

As you are likely aware, I have spent virtually my entire adult life studying Russia and the Soviet Union. My doctorate is in the history of your country. I spent more than a quarter century in the same career field as President Putin. Although he is considerably younger than I am, clearly his career has been more successful.

My first encounter with the Soviet military was 36 years ago, in 1971, when I was assigned to Vienna, Austria as an intelligence specialist to the American delegation negotiating limits on strategic armaments with the Soviet Union. While historians can and do debate whether those negotiations actually limited the number of weapons on either side, there is little debate that the dialogue between our two countries on strategic issues—doctrine, strategy, numbers, and more—made a major contribution to avoiding misunderstandings, miscalculations, and mistakes that might have resulted in a nuclear war.

I was also in Vienna in 1979 when the SALT II treaty was signed, and there for the first time met some of the top leaders of the Soviet Union—General Secretary Brezhnev, Foreign Minister Gromyko, Defense Minister Ustinov, and Chief of the General Staff Marshal Ogarkov.

In 1987, as Deputy Director of CIA, I met in Washington with Vladimir Kryuchkov, then head of the KGB's First Chief Directorate—the first summit meeting ever held between the leadership of the KGB and the CIA, a dialogue intended—like the strategic arms dialogue—to prevent mistakes and misunderstandings and to offer an opportunity to clear the air. It was a cordial, but frank, conversation. For example, when I complained to Kryuchkov about all the listening devices we had taken out of our new embassy here in Moscow, he asked me if I wanted to see the warehouse where they had all the listening devices taken out of the USSR's new embassy in Washington. Kryuchkov and I would meet next in 1989 here in Moscow, after he became head of the KGB—my first visit to the country I had studied so long.

I recall that visit well. It was May of 1989. During the trip, I had been warned that my room at Spaso House, the U.S. Ambassador's residence, would probably be bugged by the KGB. As I prepared to go to bed, I said aloud, for the benefit of whoever might be listening, that:

- I would be going right to sleep—immediately;
- I had no companionship planned for the evening; and
- Thus whoever was listening could take the rest of the night off.

I thought I heard a chuckle, but undoubtedly it was only in my imagination.

During that same 1989 trip I met General Secretary Gorbachev for the first time. As you may know, during that period I had the reputation of being a pessimist as to whether his economic reforms here would work in the later half of the 1980s—a pessimism ultimately justified by events. Thus, I was not always the favorite of the Soviet leadership or America's State Department. I recall a plenary session where Mr. Gorbachev expressed his hope that relations between our two countries might improve to the point where "Mr. Gates would be put out of a job." As it turned out, it was he that would be put out of a job first.

My last visit here prior to becoming Secretary of Defense was as Director of CIA in the fall of 1992, 15 years ago. With the Cold War by then history, I came to explore with my Russian counterpart—the head of the Russian Foreign Intelligence Service, Evgeniy Primakov—opportunities for the American and Russian services to begin to cooperate in addressing common threats in a post Cold War world: terrorism, proliferation of weapons of mass destruction, global organized crime, narcotics trafficking, and more. No longer enemies, we began to look for ways in which we could cooperate and be partners.

To symbolize the end of an era, and I hoped, the beginning of a new one, I made a special presentation to President Yeltsin during that visit. During the mid-1970s, the United States had made a major effort to recover a Soviet ballistic missile submarine that had sunk deep in the Pacific Ocean several years before in the hope of an intelligence treasure trove. We did recover part of the submarine, including the remains of six Soviet sailors. Nearly twenty years later, I presented to President Yeltsin the Soviet naval flag with which we had shrouded the coffins of the six Soviet sailors, along with a video tape of their burial at sea, complete with prayers in Russian for the dead and the playing of the Soviet national anthem—at the height of the Cold War, a dignified and respectful burial at sea of six brave adversaries.

Much has happened around the world since my visit as head of CIA 15 years ago. Much has changed. But one thing that has not changed is my belief that in this new era, Russia and America face certain security challenges in common and that there are opportunities for us to cooperate in meeting those challenges. I was eager to begin that endeavor 15 years ago. I am still eager to make that effort today.

To that end, the Secretary of State, Dr. Rice, and I have been in Moscow this week. We appreciated the opportunity to have a full and frank exchange of views with President Putin, Foreign Minister Lavrov, and Defense Minister Serdyukov. These conversations will continue, and I firmly believe they will continue to benefit the relationship and standing of both our nations.

Now, I would like to outline some thoughts on the challenges facing the U.S. military as we continue to transition to a posture more appropriate for the 21st century.

Many of these questions may be of particular interest to you as Russia seeks to modernize and professionalize its own armed services. Our nations face many of the same threats and also many of the same dilemmas when it comes to preparing our militaries for the future. I hope that I may offer a few ideas as a starting point for discussions both within your ranks, and between Russian and U.S. counterparts.

The term most often used to describe recent military developments is "transformation," or the "Revolution in Military Affairs." These expressions entered the lexicon of the U.S. military a number of years ago as ways to describe the potential for new technologies to fundamentally alter the nature of war.

What is less well known, especially in America, is that much of the original thinking on these matters was done by the Soviet military as far back as the 1970s, when officers wrote about what was then called a "Military Technical Revolution"—how to use new sensors, reconnaissance, and battle-management systems to gain an edge on the battlefield.

This work continued through the next decade, when Marshal Ogarkov, the Chief of the Soviet General Staff, envisioned a scenario where conventional systems could be as effective and dangerous as weapons of mass destruction, owing to the gains made in precision, information technology, and communications.

During the period following the Vietnam War, the U.S. military underwent its own period of transition and evolution—involving technology as well as organization and strategy. The driving force, as we all know, was the ever-elusive search for a decisive edge against our Cold War opponent. These innovations bore fruit for America's armed forces—at least with regard to conventional warfare –in the first Gulf War against Iraq in 1991.

Technological advances continued in the decade that followed, even as the U.S. military suffered an identity crisis of sorts. No longer were we seeking to deter any single nation-state. No longer was the mission geared to one nation alone. Our military was instead tasked with new types of peace-keeping and humanitarian missions in places like Somalia, Haiti, and the Balkans.

Since September 11, the U.S. military has been confronted with new missions in Iraq and Afghanistan—where initially quick conventional victories have given way to long, complex, and grinding campaigns against violent, adaptive insurgencies.

When these conflicts began, the U.S. military was, for the most part, a smaller version of the force that had been organized, trained and equipped to do battle with the Soviet Union. In light of the lessons of Afghanistan and Iraq, the U.S. military has undertaken a number of reforms to prepare for what will likely be a generational campaign against violent Islamic extremists. A few examples:

- We no longer have the luxury of months to deploy, assemble, and prepare as was the case in 1991. So our ground forces, and in particular the U.S. Army, are becoming more agile and expeditionary.

- As the U.S. military became a more professional force—conscription ended in 1973—the welfare of military families took on greater importance—in morale, retention and other areas. So we've learned to do more to support and sustain the families of our servicemen and women.

- Similarly, the role of reserve forces has changed—from that of a "strategic reserve" to be mobilized in the event of a third World War to an "operational reserve" which is now an integral part of every mission.

The very idea of transformation has gone well beyond its original technological associations. It now stands for a process of constant evaluation, adaptation, and change. It is an ongoing process, and more needs to be done.

Looking forward, one of the most important roles for our military will be less direct combat and more a matter of helping the security forces of partner nations defeat extremists within their own borders. So we're thinking about how best to incorporate those capabilities into our Armed Forces in a way that does not detract from their war-fighting responsibilities.

Furthermore, the struggle against terrorists is a campaign that requires the full use of all elements of national power—diplomatic, economic, political, as well as military—in close cooperation with allies and partners. The Defense Department is seeking better ways to integrate with other agencies of the U.S. government. As I'm sure many of you know, bureaucratic inertia is always a source of frustration. But we have had some success. One example is the concept of Provincial Reconstruction Teams that have been used in Afghanistan, and more recently in Iraq. These units bring together soldiers from the U.S. and other nations with experts in agriculture, reconstruction, and other fields.

We are also trying to be more creative when it comes to using the talents and capabilities of our country as a whole, and not just our government. A few months ago, I expanded a program to assign civilian anthropologists and social scientists to each of our deploying brigades. The cost of the program was miniscule compared to a new weapons system or a traditional military operation—but the positive impact can be significant, as a number of commanders in the war zone can attest.

These kinds of changes are difficult for any large institution that has been doing things a certain way for a long time. They are difficult in any era. When I was CIA director in 1992, the agency was in the midst of a dramatic transition away from a focus on the Soviet Union and toward other threats. It meant reassigning and retraining thousands of people. It meant reexamining long-held assumptions and ways of doing business. And at that time, it meant doing all this while the CIA's budget and personnel were being drastically reduced—a circumstance I know leaders of the Russian military had to cope with in the past.

Though our nations and our militaries are in very different places today, we do face many of the same challenges. One of the reasons I chose this topic for presentation is that I wanted to provide an outline of what U.S. defense leaders are thinking about as we consider the future. I hope this will go some way toward creating a climate of trust and transparency as our countries take on some of the thornier geopolitical issues of the day—and subjects for further discussion in the future.

I would just close on a personal note. A number of months ago, I had an opportunity to stand on the shores of Normandy and pay homage to the Americans who had lost their lives on that World War II battlefield. A month before, President Putin spoke at a ceremony marking the anniversary of the end of the Great Patriotic War, in which more than 20 million citizens of the Soviet Union perished. We were two old Cold Warriors—and plain-spoken career spies—honoring the deeds of a bygone era, deeds shared by an America and a Russia that allied against a common enemy, together. We were honoring the mass heroism of rank-and-file soldiers, the unconquerable spirit of everyday citizens, the tragedy of great nations.

President Putin said in his remarks that "We have a duty to remember that the causes of any war lie above all in the mistakes and miscalculations of peacetime and that these causes have their roots in an ideology of confrontation and extremism."

No nation suffered more from the last century's carnage and miscalculations than Russia. And today, arguably no nation stands to gain more from this century's possibilities. We are prepared to work with Russia—and with the Russian military—to try to turn possibility into reality for the peoples of both nations.

## Jewish Institute for National Security Affairs

*Arlington, VA, Monday, October 15, 2007*

THANK you Bob. Thank you for that introduction. Thank you for inviting me here tonight. And thank you for this honor. Above all thank you to all those recognized with the grateful nation award. You are the best and we all owe you and in all sincerity we are all humbled by you. The truth is, compared to you, I don't know what the hell I am doing up here.

Well you know relations between the United States and Israel have been very close for a long time and there've been moments of drama and moments of humor. One of the latter was the time that President Nixon met with Israeli Prime Minister Golda Meir just after he appointed Henry Kissinger as Secretary of State. And Mrs. Meir was accompanied by the very erudite Abba Eban. And Nixon turned to the Prime Minister and said "Just think we now both have Jewish foreign ministers." And Golda Meir looked at Nixon and said, "Yes, but mine speaks English."

The reception this evening and wine with the dinner and all of that reminds me of the risks of officials drinking in public. Some years ago, a European foreign minister who shall remain nameless, who was a notoriously heavy drinker, was on a trip to South America and he showed up at a reception in Peru and he was quite drunk. And there was music playing and he invited a passing guest to dance. And the guest somewhat haughtily replied, "First, sir, you are drunk. Second, this is not a waltz, it is the Peruvian national anthem. Third, I am not a woman I am the Cardinal Archbishop of Lima."

There's another example of this and it was passed along to me by Bob Strauss a number of years ago and it was the story of a long-winded after-dinner speaker. There'd been a long reception and lots of cocktails and then lots of wine with dinner. The speaker was at one of those table-set podiums, so there were people seated to both sides. And the speaker got up and droned on and on. And finally a drunken guest seated to the right of the speaker got fed up, picked up an empty wine bottle and swung it at the speaker and missed, and hit the chairman of the event who was seated on the left of the speaker who fell to the ground bleeding profusely. The drunk got down on his hands and knees crawled over to the chairman of the event to apologize. The chairman opened one eye and said, "hit me again I can still hear the son of a bitch."

So I promise not to go on too long. Although I think I am out of range.

For more than three decades, the Jewish Institute for National Security Affairs has been a strong supporter and defender of America's military might—and of America's long-standing friendship with Israel. I thank you all for your contribution to the debate over our nation's foreign policy.

I must say that it is humbling to receive an award named after Scoop Jackson, one of the great senators of modern times. It is more humbling when one considers the long line of distinguished men and women who have stood before this organization, in receipt of this award.

For those of us who had the pleasure of working with Scoop Jackson on national-security issues, even if only for a couple of years in my case, his was a voice and a vote that could always be counted on to defend America's interests abroad. He had a profoundly realistic understanding of human nature and how the world worked. He was fond of saying that he was neither a hawk nor a dove—he just didn't want his country to be a pigeon.

When it came to national security, his was the consummate bipartisan voice. He always said that, in defense matters, "the best politics is no politics." He was an outspoken and implacable critic of anti-Semitism, both at home and abroad. He understood that in many places Jews lived under oppressive conditions and governments. So he put forward legislation to pressure other nations, particularly the Soviet Union, to allow their citizens, and particularly Jews, to immigrate to friendlier nations, often Israel. He personally intervened on behalf of Natan Sharansky and many others trapped behind the Iron Curtain.

On a fundamental level, I think he felt something that many presidents have felt—from John Adams to Abraham Lincoln to Harry Truman to Ronald Reagan and George Bush. Like all of them, he believed that the Jewish people deserved a home in their ancient lands. And he, like many, made it his duty to protect and preserve that nation.

Since the creation of Israel, and even before, there has been no small amount of discussion about what our relationship with the Jewish state should be. And no small amount of ink has been spilled about what our interests are—whether strategic, political, moral, or some combination thereof.

Perhaps the best summary of our reasons for supporting Israel may also be the most succinct. In 1967, Soviet premier Aleksei Kosygin asked Lyndon Johnson why he supported the Jewish state, even though its population represented only a fraction of the entire Middle East. And President Johnson replied: Because it is right.

It was right to stand by Israel during its darkest hours when it fought for its very survival. And today, with the new threats and challenges our nation faces in the region, it is even more important to maintain and bolster our partnership.

My personal experiences with Israel date back almost 35 years, and begin with one of my most embarrassing moments as an intelligence officer. I was in Geneva, Switzerland as an intelligence adviser to the U.S. strategic arms delegation in the fall of 1973. I was giving Ambassador Paul Nitze his morning intelligence briefing, and his eye was caught by one item in particular—the CIA's analysis that Egypt would not attack Israel. Nitze asked me if I spoke French. I said no. He asked if I listened to the radio. I said no. He said, "Well, if you listened to the radio and understood French you would have known before you came in here that Egypt has already attacked Israel." The Yom Kippur War had started that morning.

My first visit to Israel was nearly 30 years ago, in the final phase of the Camp David process. I was Zbigniew Brzezinski's special assistant. We finished our work at the King David Hotel about 2 in the morning and the then-Deputy Executive Secretary of the Department of State, Frank Wisner, asked if I wanted to go for a walk. And so, my first walking tour of ancient Jerusalem was in the middle of the night. It was one of the most profoundly religious experiences of my life. The shops all shuttered, no tourists—in fact, not a soul in sight. I was transported in time back two thousand years. No distractions to interrupt the experience, just the sound of our shoes on the stones. I will never forget it. It's probably a good thing, too, because I expect if I tried to do that today, the legions of security—mine and Israeli—would probably detract from the experience.

I would visit Israel on a number of occasions as Deputy Director and Director of CIA, with the opportunity to meet and work with then-Defense Minister Yitzhak Rabin, Prime Minister Shamir, several heads of Mossad and Shin Bet, and then-Chief of Staff Ehud Barak. Thus, it is not by accident that one of my first foreign trips included Israel and a meeting with Prime Minister Olmert and others. At the time of my visit, it had been more than six years since an American Secretary of Defense had visited Israel. And, while it has been many years since I have seen my friend, Ehud Barak, we have talked on the telephone and I look forward to seeing him tomorrow. You know, we old Kremlinologists pay a lot of attention to the little signs and gestures that matter a lot—I will share with you that, ten months into this position, tomorrow night I will host my first dinner for a foreign minister of defense—for Ehud Barak.

Well, enough reminiscing. As one looks around the broader Middle East today—wars in Iraq and Afghanistan, an ambitious and fanatical theocracy in Iran, a malignant terrorism, and more—some might believe that our bond with Israel is more a hindrance to stability in the region, an irritant for already tense relationships.

In some respects, I think that view is borne of frustration, the outcome of a decades-long attempt to bring a lasting peace to Israel and the Middle East. And yet, despite ongoing violence perpetrated by militant, jihadist organizations—as well as the strident posturing of Iran—progress has in fact been made.

Israel's existence is no longer threatened or challenged by the armies or air forces of its immediate neighbors. The nation is recognized and at peace with Egypt and Jordan—relations that, while tenuous at times, are nonetheless a far cry from 1948, 1967, and 1973. Iraq no longer has a belligerent government, and no longer holds Israel's safety hostage to deter actions by other nations. And, just a few months ago, the Arab League sent its first-ever envoy to Israel.

I point this out because, as intractable as the situation may sometimes seem, the perspective of history does offer some degree of hope for relations among the governments of the Middle East. Some degree of hope that Israel will not forever be watching its back.

History also has something to say about the tense situation with Iran, an issue of great importance not only to this organization, but to the entire world.

I remember back to November 1, 1979, when then-National Security Advisor Brzezinski was in Algiers representing the U.S. at the 25th anniversary of the Algerian Revolution. As his special assistant, I was along, and so was his head of Congressional Liaison, a woman named Madeleine Albright. While we were there, the Iranian delegation asked to meet with Brzezinski. I accompanied him as the note taker. Zbig offered the Iranians—their Prime Minister and Defense and Foreign Ministers—recognition of their revolution, continuation of their partnership that had existed under the Shah—including military assistance to the new government, and focus on a common foe to Iran's north—the Soviet Union. They weren't interested. They only wanted us to give them the dying Shah. Brzezinski refused, finally saying that to return the Shah would be incompatible with our national honor. That ended the meeting. Three days later came word that our embassy in Tehran had been seized, and two weeks after that, the prime minister and defense and foreign ministers with whom we had met were out of their jobs and/or in jail. Thus began my now 28-year-long quest for the elusive Iranian moderate.

We should have no illusions about the nature of this regime or its leaders—about their designs for their nuclear program, their willingness to live up to their rhetoric, their intentions for Iraq, or their ambitions in the Gulf.

This Administration is keenly aware of the threats posed by Iran. It is also keenly aware of the challenges we and our allies face with a regime that seems increasingly willing to act contrary to its own national interests. With a government of this nature, only a united front of nations will be able to exert enough pressure to make Iran abandon its nuclear aspirations—a source of anxiety and instability in the region. Our allies must work together on robust, far-reaching, and strongly enforced economic sanctions. We must exert pressure in the diplomatic and political arenas as well. And, as the President has said, with this regime, we must also keep all options on the table.

But, obviously, instability in the region is not just driven by state actors. The recent history of the Middle East has demonstrated the lethality and persistence of armed militias and movements that have no allegiance to any government, only to death and destruction and chaos. Where extremists have seized and controlled territory—in western Iraq or eastern Afghanistan, for example—the result has been misery, and poverty, and fear. The future they promise is a joyless existence—personified not by piety or virtue, but by the executioner and the suicide bomber. Symbolized by men kneeling not in prayer before their god, but kneeling and waiting for the executioner's sword.

The United States and many of our allies, the prospect of terrorism on a large or prolonged scale is a relatively new concept, one that we are just beginning to appreciate. For Israel, however, it is something that dates back many years.

Despite many tactical successes, overall strategic success against violent extremism has been elusive. With the extent of the jihadist movement, with its breadth and numbers, even the most effective counterterrorism tactics can only reduce the number and lethality of attacks. Total elimination is infinitely more complex, part of an ideological struggle between the forces of moderation and extremism. It is a struggle currently playing out in Iraq and elsewhere in the Middle East.

I have in the past few months outlined a number of consequences of American failure in Iraq—consequences that would affect and potentially destabilize all the nations of the Middle East. One of the consequences of greatest concern is ideological in nature.

The jihadist movement draw their support largely from perception—not necessarily reality.

Consider the prevalent myth that jihadists were solely responsible for bringing the Soviet empire to its knees in Afghanistan in the 1980s. The reality is far less romantic. The very diverse Afghan resistance was only able to be as effective as it was because of massive support from the United States and likeminded nations—support in weapons, in strategy, in dollars, in sanctuary. Without those things, the insurgency may have ground on, but at an acceptable cost to the Soviet Union. And yet, Osama bin Laden, not yet a key player in Afghanistan at that point, has embraced, co-opted, and severely distorted this history for his own ends.

A similar phenomenon could happen in Iraq if we pull out quickly, leaving chaos in our wake. It is not at all a stretch to imagine an all-out propaganda campaign in which the jihadists portray themselves as defeating not one, but two superpowers. That would surely dramatically embolden an entire generation of Islamic extremists, and encourage countless others to join their ranks and wage war on our allies and our interests in the region, in Europe, and ultimately here at home.

The challenges in Iraq clearly are steep. Even so, I think there are reasons to be hopeful.

Over the last few months, there have been a series of public expectations about what the President should do in Iraq. It started with asking him to say he would draw down our forces; then it was a date to begin the drawdown of forces; then it was a timetable for the drawdown of those forces; and then it was asking him to state that there would be a change of mission.

The President has moved on all of those. He has announced that there would be a drawdown; he announced when it would begin—and that was a few weeks ago; he accepted General Petraeus's schedule through next July with a review in March to see what to do beyond July, a conditions-based timetable; and the President announced that this December would mark the beginning of a transition of mission.

Most people now are focused on whether the drawdown is swift enough, whether the timelines should be binding, and so forth. It is not, it seems to me, a debate, however, about the overall trajectory of the President's plan. And I think that's is an important distinction, and promising for the creation of a bipartisan consensus for the future. I would only add that I hope those who have alleged that the views of our generals were neglected at the start of this war will not now dismiss the unanimous recommendations of our generals for the next steps.

I have on many occasions spoken of the need to get the next phase in Iraq right. How we got to this point, whether and what kind of mistakes were made, will undoubtedly be the subject of historical analysis for a long time to come. Right now though, members of both parties are realizing the full extent of the challenges we face—how dangerous a failed state in Iraq and an ascendant Al Qaeda would be, not just in the short-term, but for decades to come. And, despite the sometimes acrimonious debate, I believe that members of both parties are slowly coming to the same conclusions about our future course in Iraq—even if they disagree on dates and details.

Progress toward peace in the Middle East, whether between nations or between peoples, has always been to the advantage of the United States—both in terms of our values and in terms of our interests.

But if we are ever to see a lasting peace, we must keep our objectives foremost in mind:

- A unified and stable Iraq;
- A just and comprehensive peace between the Israeli and Palestinian people, including, as the President has said, a two state solution;
- An Iran that does not attempt to dominate the region by subverting its neighbors, by building nuclear weapons, or by holding Israel hostage with the threat of attack; and
- A reversal of the growth and influence of extremist networks and sectarian militia organizations that have become, in the words of our former theater commander, "the curse of the region."

I'd like to close by taking a step back in time to a period of great consequence for Israel and the formation of the modern Middle East.

During the late 1890s, the Ottoman Empire was under great stress, the Balfour Declaration still more than 20 years off, and change was in the air. Both Zionism and Arab nationalism were still nascent movements, but their roots nonetheless had begun to take hold in the Middle East and elsewhere.

With tensions rising, Theodor Herzl responded to one of his critics in Jerusalem. He wrote, "As a people, [the Jews] have long lost the taste for war. They are ... fully content if left in peace."

I think that is a fitting description of all the mature democracies in the world. We have no taste for war, no taste for the destruction and devastation that it creates. We are content to live in peace.

But if we are not left in peace, if our security is challenged, we also know that there may be times when we have to defend in no uncertain terms our interests and our liberties. Some in the Middle East and the Persian Gulf and have said the United States is weak because of our engagement in Iraq and Afghanistan. Restraint should never be confused with weakness.

And let no one, friend or foe, ever forget the words of Dwight Eisenhower, who wrote before World War II, "... beware the fury of an aroused democracy," and who wrote after the war, "It is a grievous error to forget for one second the might and power of this great republic."

We will do our duty to our people. We will do our duty to our allies and friends. And we will do our duty to posterity. Just as Scoop Jackson would have us do.

Thank you.

# STRATCOM Assumption of Command

*Offutt Air Force Base, NE, Wednesday, October 17, 2007*

GOOD morning. It's a pleasure to be here in Omaha and on Offutt Air Force Base. Of course, it's always good to be at least 1,160 miles and at least one time zone away from Washington.

It is an honor and a pleasure to leave the beltway and come to a place where people truly deserve a pat on the back and much more.

I would like to welcome the family of Gen. Chilton, especially his wife, Col. Cathy Chilton, their four daughters, her parents, General and Mrs. Dreyer. Also, I'd like to thank Governor Heineman and our many distinguished guests for coming. Since I came to this job from the presidency of Texas A&M, the governor and I had a word or two about this coming Saturday. Both of us are proud, but not boastful. On behalf of the President and our colleagues in the Congress, I would also like to express my greetings and appreciation to the men and women of the U.S. Strategic Command.

Visiting this organization carries a special meaning for me, having spent most of my time as an Air Force lieutenant some 40 years ago, serving in the old Strategic Air Command—then responsible for America's nuclear deterrent. Though a good deal has changed in the decades since, the proud tradition of safeguarding our country from the most destructive threats continues in this organization today.

Your determination and skill have helped propel this vital mission forward. It is a young mission for the most part, particularly in the area of space and cyberspace. In addition to bringing with you the great traditions of your various Services, you add to the spirit of our nation's parent pioneers. And we are grateful for your service.

General Kevin "Chili" Chilly Chilton is someone who has blazed new trails for much of his career:
- As a test-pilot on the F-4 and F-15;
- In a series of key staff and command assignments, including Air Force Space Command; and
- Of course, as an astronaut who flew on three space shuttle missions.

I believe General Chilton has the right set of skills and experience to lead this Command at this time.

Today, America faces a range of security challenges, including an unrelenting extremist enemy opposed to our way of life and determined to do us harm—an enemy not only found in Iraq and Afghanistan. The potential of such terrorist groups gaining control of weapons of mass destruction is arguably the greatest threat facing our nation today.

In addition to stopping the proliferation of dangerous materials, a key part of our strategy for combating this threat—and a major responsibility of this Command—involves space-based capabilities.

It is through space that we can monitor the weapons we already know exist.

It is through space that we can track adversaries attempting to acquire these weapons, and then do something about it.

It is through space that our troops and our leadership monitor the battlefield and communicate with each other.

Therefore, it is space that we must protect, especially as we expand its use.

The importance of maintaining unfettered access to space was reinforced earlier this year, when China successfully tested an anti-satellite weapon. This test and other developments show that our own near-earth satellites are vulnerable, and must be protected.

General Chilton, we look to you for the way ahead during these turbulent times. As former director of CIA, I understand the importance of strong intelligence-gathering systems, a space-based system often. I also understand the need to analyze that intelligence and to plan our response.

I am confident that you will be able to tackle the challenges ahead, and build on the accomplishments of this command under Gen. Cartwright.

During his 3-year stint, General Cartwright dramatically flattened the organization, which sped up the flow of information throughout STRATCOM. Another of his initiatives was the Global Innovation and Strategy Center which gathers a group of outside experts and brings them in to concentrate for a limited period of time on a specific issue. They zero in on the problem, brainstorm from their various professional points of view, develop recommendations, and then disperse, leaving STRATCOM leaders to do the rest. When I heard about these programs, not to mention all of his other accomplishments, I knew we needed Hoss Cartwright back at the Pentagon.

I also knew that it would be hard to find someone who could fill his shoes. And I believe we have found that person in General Chilton.

General, I look forward to seeing the great things you will do here with this able staff.

As the great conductor Leonard Bernstein once said, "To achieve great things, two things are needed: a plan, and not quite enough time." I'm sure both apply here.

Good luck to you, and I wish you and Cathy the best.

# Conference of European Armies

*Heidelberg, Germany, Thursday, October 25, 2007*

THANK you, General McKiernan. General McKiernan and I went to the same college, though I confess I graduated some years earlier than he did.

I am pleased to be part of the 15th annual Conference of European Armies. As a former second lieutenant in the Air Force, I never dreamed I'd be standing before so many top army generals, much less at a conference on "Transforming Land Forces in the 21st Century." And it is an honor to be here.

The defense of this continent has been a consuming goal for me for most of my professional life—more than four decades, under multiple administrations of both American political parties. I've seen firsthand the powerful synergy that comes from free nations pulling together to defend our shared values and interests. Since the collapse of the Soviet Union and since becoming Secretary of Defense, I have spent a good deal of time considering how the alliance must evolve in order to remain vital and relevant in a new era.

To explore such questions, you have come together at what was once the geo-strategic crosshairs of the Cold War, just over 100 miles from the Fulda Gap. No troops crossed that gap. A collective posture of strength and resolve among allies of four decades prevented a violent showdown between NATO and the Warsaw pact from ever taking place.

One key test of wills took place in the 1970s, when the Soviet Union deployed SS-20 ballistic missiles in Europe. In response, the Alliance agreed to strengthen its conventional and nuclear capabilities. At the time, I was working on the National Security Council under President Carter.

In December 1978, two colleagues and I had the good fortune to do advance planning for the NATO summit that would address these concerns. I say "good fortune" because the summit was being held on the French island of Guadeloupe, with its easy ocean breezes and mild Caribbean sun. When we arrived at the hotel, our French hosts were waiting for us at a small pavilion on the beach. We stepped onto the white sand feeling quite awkward in our neck-ties and suits—but that would become the least of my worries. It turned out the pavilion where we were conducting the planning meeting was located in the middle of a topless beach. Luckily, the Guadeloupe summit later went off without a problem. But, I must admit, my notes from that seaside meeting were quite incomplete due to nearby distractions.

In all seriousness, the Guadeloupe summit addressed one of the most critical issues in NATO's history: the decision to deploy cruise and Pershing missiles in Europe. This policy decision was politically difficult for many Allies, but it ultimately set the stage for deep reductions in nuclear arms and the end of the Cold War.

Those kinds of decisions required frank discussion, firm resolve, and bold action that transcended the differences among our nations. Although NATO was originally created to oppose Soviet communism, its guiding principle was broad and deep and still holds true—to safeguard the freedom and security of all its members for generations to come.

In the next few minutes, before taking several questions, I would like to share some of my concerns about NATO's current operations and capabilities.

We meet at a time when the Alliance that never fired a shot in the Cold War now conducts six missions on three continents. NATO forces are deployed in Afghanistan—thousands of miles beyond Europe's borders—and are engaged in significant ground combat for the first time. They are helping the Afghan people rebuild and secure their country so it will never again harbor terrorists plotting to attack the west or have a government that so ruthlessly and savagely oppresses its own people. The International Security Assistance Force has nearly 40,000 boots on the ground backed by almost 40 allied and partner nations.

Approximately one year ago, NATO assumed primary responsibility for security in Afghanistan, and it is worth taking stock of what has been accomplished:

- Today the Alliance leads 25 Provincial Reconstruction Teams—digging wells, building schools, and paving roads;
- Decisive military operations by some allies thwarted the Taliban's "Spring Offensive"; and
- The Afghan Army is 47,000-strong with representatives from every major ethnic group. It is growing steadily, but to reach the goal of 70,000 by the end of next year urgently demands more mentoring and liaison teams than we have now.

ISAF's accomplishments are real and tangible to the citizens of Afghanistan. Someone once wrote that we must all "plant trees we will never get to sit under." The NATO alliance will indeed reap a lasting bounty from its efforts in Afghanistan—if we can muster the resolve and will to get the job done.

Said differently, our progress in Afghanistan is real but it is fragile. At this time, many allies are unwilling to share the risks, commit the resources, and follow through on collective commitments to this mission and to each other. As a result, we risk allowing what has been achieved in Afghanistan to slip away.

Earlier today you attended a session addressing the issue of caveats. This is an important topic that warrants further mention, as it is symptomatic of a deeper challenge facing NATO. Imagine a game of chess in which one player enjoys full liberty of motion, while another may move rook, bishop, and queen only a single space in a single direction like a pawn. One player is clearly handicapped. Similarly, restrictions placed on what a given nation's forces can do and where they can go put this Alliance at a sizable disadvantage.

I recognize the need for political oversight of deployed ground forces and—as someone who has signed far too many condolence letters this year— I also understand the desire to avoid casualties. I also acknowledge the fact that the political and economic landscape for each of you in this room is different. Europe is far from monolithic. I recall former U.S. Secretary of State Henry Kissinger's lament: one could not pick up a phone and call "Europe" when facing an international crisis.

While there will be nuances particular to each country's rules of engagement, the "strings" attached to one nation's forces unfairly burden others, and have done real harm in Afghanistan. As you know—better than most people—brothers in arms achieve victory only when all march in step toward the sound of the guns. To that end, I'm asking for your help to make caveats in NATO operations—wherever they are—as benign as possible, and better yet, to convince your national leaders to lift restrictions on field commanders that impede their ability to succeed in critical missions.

I am also concerned that the Alliance is not transforming quickly enough to deal with the realities of the 21st century. The U.S. military knows firsthand that real and lasting change in any large organization is difficult, messy, and often slow—though we have made considerable headway in recent years in the way our forces are organized, trained, and equipped. Achieving fundamental change is altogether more difficult when it is done as a cooperative process among many nations and military establishments.

The alliance is evolving but must continue to shift from a reactive, static posture to a more proactive, expeditionary one. We must continue making institutional reforms, such as streamlining headquarters, expanding information sharing, and improving command and control. As our experiences in Bosnia, Kosovo, and Afghanistan have shown, the type of military operations that will require NATO forces has dramatically changed in the 21st century. Even the most advanced weaponry is no substitute for "boots on the ground" helping to quell ethnic conflicts, fight terrorists, and rebuild communities.

Doing all of these things requires a willingness to alter long standing and comfortable ways of operating. Building these new capabilities takes real political will, as does modernizing and transforming a military, which is never cheap and competes with other domestic priorities.

At the summit at Riga last November, we declared the NATO Response Force fully operational. Since then, we have reaffirmed the NRF's twin roles: to catalyze transformation, and to be availablefor unforeseen needs—in ongoing operations as well as new crises. For the NRF to be successful, it will require full allied political support—both in terms of pledges to the force and agreement that the force just won't sit on the shelves. Progress in these areas—and on other allied contributions—are factors that will help determine what future U.S. contributions will be.

This leads me to the final topic I'd like to address before closing—indeed the one that underlies most of the challenges we face in this Alliance—and that is commitment.

This alliance is not a "paper membership" or a "talk shop." It is a military alliance and one with serious real world obligations. Over the years, Europe and the Alliance have been characterized in various ways: "East" and "West," the "free world" and "those behind the Iron Curtain," even "new" and "old." As I said earlier this year in Munich, my characterization of the membership of the Alliance is a practical one—a realist's view, perhaps. Simply stated, there are those members who fulfill their commitments, and those who do not.

In a broad sense, commitment implies a willingness to accept measured, calculated risks and to fulfill pledges. For example, a widely recognized benchmark is for Allies to spend 2 percent or more of GDP on defense. Yet currently, only 6 out of 26 NATO members have met that goal. We must live up to the pledges we agreed upon at Riga and reverse the slide in Allied defense spending.

Something tells me those of you wearing your nation's uniform have sought bigger defense budgets; at the same time, you likely know the difficulty of making that happen. One of my first duties when I became Secretary of Defense was to present the Department's base budget and war requests for the next fiscal year. I'm here to tell you that you haven't lived until you've gone before your nation's elected representatives and asked for nearly $750 billion dollars from the taxpayers' wallets for one year's expenditures. That was, as they say, a rough day at the office.

Many of you know General Jim Jones, former Supreme Allied Commander in Europe. Jim made a provocative and compelling point about a year and a half ago, when he said it was easy to sail NATO's navies and fly NATO's planes, but a challenge to deploy NATO's armies.

As it stands today, non-U.S. NATO nations have more than 2 million men and women in uniform, yet we struggle to maintain 23,000 non-U.S. troops in Afghanistan. This is partly a function of how NATO militaries are organized, and partly a matter of resources—but it is mostly a matter of will and commitment. The same is true for equipment and other resources. Consider that earlier this year the U.S. extended its Aviation Bridging Force in Afghanistan in Kandahar because the mightiest and wealthiest military alliance in the history of the world was unable to produce 16 helicopters needed by the ISAF commander. Sixteen.

Meeting commitments means assuming some level of risk and asserting the political will necessary to deploy armed forces beyond one's borders—fully manned and equipped, and without restrictions that undermine the mission. In Afghanistan, a handful of allies are paying the price and bearing the burdens of allies to create the secure environment necessary for economic development, building civic institutions, and establishing the rule of law. The failure to meet commitments puts the Afghan mission—and with it, the credibility of NATO—at real risk. If an alliance of the world's greatest democracies cannot summon the will to get the job done in a mission that we agree is morally just and vital to our security, then our citizens may begin to question both the worth of the mission and the utility of the 60-year-old transatlantic security project itself.

In closing, I'm sure many in this room have watched the extraordinary trajectory of our trans-Atlantic partnership arc from detente in the 1970s to the dissolution of the Warsaw Pact in the 1990s. Near-confrontations between the superpowers took place, and relations between the allies were not without stress and strain. But our partnership weathered these difficulties. We made decisions with sharp edges, guided always by our shared past and common values. At the Prague summit nearly five years ago, President Bush stated, "we are tied to Europe by the wars of liberty we have fought and won together. We are joined by broad ties of trade. And America is bound to Europe by the deepest convictions of our common culture—our belief in the dignity of every life, and our belief in the power of conscience to move history."

And so we must continue building partnerships capable of flexing military might when and where needed—a prerequisite against an adaptable, transnational enemy. Alexander the great said, "Remember—upon the conduct of each depends the fate of all." We triumphed in the Cold War because of our ability to surmount individual differences and unite against a common foe. The stakes today are just as high.

Thank you.

## George Bush Award for Excellence in Public Service

*College Station, TX, Friday, October 26, 2007*

HOWDY! I love saying that. I tried saying it yesterday morning at the NATO defense ministers meeting in the Netherlands. It didn't get quite the same response. Mr. President, Mrs. Bush, ladies and gentlemen, Aggies, I can't tell you how wonderful it is to be back in Aggieland. Mr. President, I am deeply moved and honored by this award, especially because it bears your name—one of the most distinguished public servants in American history.

One of the great privileges of my life was to be at this president's side as he provided inspired leadership to a world that in a span of less than 36 months, experienced the liberation of Eastern Europe, the reunification of Germany and NATO, the victory of the West in the Cold War, the collapse of the Soviet Union, and the first Gulf War. As I wrote in my book over 10 years ago: "The imagination reels at the thought of a less experienced and skilled president trying to exploit the liberation of Eastern Europe or dealing with the final crisis and death throes of the Russian and Soviet empire …. As the communist bloc was disintegrating, it was George H.W. Bush's skilled, yet quiet, statecraft that made a revolutionary time seem so much less dangerous than it actually was."

Ten years after I wrote those words, I would add that I am confident that someday the world will recognize the boundless debt it owes this President Bush for his leadership during a time when, without precedent in modern history, a great and powerfully armed empire came to an end without a war. I am humbled to be his friend and to be honored by him.

Now, working for President Bush—the 41st president, now known simply as "41"—was not all drama. He is, for example, the creator of another prestigious award: the Scowcroft Award, named for his national security advisor, Lieutenant General Brent Scowcroft. This award was created by the President in 1989 to honor the American official who most ostentatiously fell asleep in a meeting with the President. This was not frivolous. Candidates were evaluated on three criteria: First, duration—how long did they sleep; second, the depth of the sleep—snoring always got you extra points; and third was quality of recovery—did one simply quietly open one's eyes and return to the meeting or jolt awake, possibly spilling something hot? General Scowcroft was, of course, the first awardee, and I might also add won many oak leaf clusters.

Some of you might remember that the cartoonist Garry Trudeau made quite a good living at the President's expense in his cartoon strip Doonesbury. The strip often featured 41's invisible other self—"President Skippy"— as an asterisk. One morning when the President stepped out of the Oval Office during a briefing, we had a photographer come in and take a picture of Scowcroft, Chief of Staff John Sununu, and me all talking and gesturing vehemently at an empty chair. We later presented a large, framed copy of the photograph to him, inscribed, "To President Skippy, from the gang that knows you best."

He loved it, and promptly jumped up out of his chair and said, "The press has to know about this." He then walked to the press room without any forewarning, and there was nearly a press riot. He showed them the picture, said there was clearly a plot against him inside the administration, and then attributed the whole idea to his spokesman, Marlin Fitzwater—a completely innocent and unwitting bystander.

I owe much to both former President Bush and to General Scowcroft, but I will never be able to repay what I owe them for introducing me to Texas A&M. Of course, the method they used, as befits a former president and head of CIA, and a national security advisor to two presidents, involved more than a little deception. I was told that being interim dean of the Bush School would be largely "honorific," a day or two a month for nine months— just until a permanent dean was found. It was a classic bait and switch con. Two weeks a month commuting from Seattle, and two years later, they finally hired a permanent dean, Lieutenant General Dick Chilcoat.

I fell in love with Texas A&M, soon after I arrived in August of 1999, but in the summer of 2001, I was headed home to the Northwest—and to my family. A&M President Ray Bowen had announced his retirement plans and several people asked me if I would be a candidate to succeed him. I asked one if he had gotten hold of a bad drug. I repeatedly said no, and went home.

Six weeks later, 19 jihadist terrorists changed all of our lives forever. Two and a half months after 9/11, on December 1, 2001, the former chairman of the A&M Board of Regents, Don Powell, made a final stab at asking me to be a candidate for President of A&M. I talked to Becky and told her that after 9/11 I felt compelled to do one more public service. I told her I never wanted to return to Washington, D.C., and so I felt I had to agree to be a candidate to lead A&M. And you know the rest of the story. And so, on August 1, 2002, after a retirement that lasted almost nine and a half years—

including two years at that "honorific" job at the Bush School—I returned to public service as President of Texas A&M.

The world turned upside down again for me almost exactly a year ago when duty—and another President Bush—called me away from here and to a different kind of public service. Just over 40 years after I took my oath and joined CIA, I was on my way back to Washington and government service.

What is this strange phenomenon called public service? Certainly many have considered it an onerous burden. Edwin Stanton, President Lincoln's Secretary of War, wrote: "There could be no greater madness than for a man to encounter what I do for anything less than motives that overleap time and look forward to eternity." Lord Cornwallis complained, "I have been obliged to say yes and exchange a life of ease and content, to encounter all the plagues and miseries of command and public station."

My favorite whine, though, came from no less a figure than Benjamin Franklin, who wrote the following to a friend who was assuming a public office: "The public is often stingy, even of its thanks, while you are sure of being censured by malevolent critics and bug-writers, who will abuse you while you are serving them, and wound your character in nameless pamphlets, thereby resembling those little dirty insects that attack us only in the dark, disturb our repose, molesting and wounding us while our sweat and blood are contributing to their subsistence."

And yet, they served.

Tonight, nearly 200,000 American men and women in uniform serve in harm's way in Iraq and Afghanistan. Every single one is a volunteer. Each has chosen to be there in the belief that they are protecting America and bringing a better life to millions long oppressed. As of yesterday, 3,385 of their comrades have died in action, in Operation Iraqi Freedom and Operation Enduring Freedom in Afghanistan—and 30,044 have been wounded in action.

And yet they serve.

Millions of other Americans have chosen careers in public service, electing to serve their fellow citizens in the belief they can help make this country and the world a better place. More than two million men and women in the active and reserve armed forces; policemen and firemen; teachers; nurses; elected and appointed officials—local, state, and national; and countless more. All too often, the pay and working conditions are challenging. All operate in the public spotlight and often find public criticism to be the reward of their labors. Many could live better pursuing other careers.

And yet they serve.

A life in public service can be challenging. But I think the challenge has less to do with often inadequate compensation and working conditions than it does with the public climate in America today. Politics was always a rough business at the top in America. After all, don't forget that Aaron Burr shot and killed Alexander Hamilton, and name-calling before, during, and after presidential elections has afflicted every president since George Washington—and even he suffered calumny. But the harsh and unforgiving environment has spread far beyond presidential politics. Even local service—on school boards, municipal councils, and the like—can expose one to ugly attacks and venomous criticism.

Any human weakness or error is pounced upon and often leads to instant ridicule in the national media or, worse, personal ruin. And, today, not only is an official subject to such scrutiny, so, too, are his or her family members, business or professional partners, and even friends. Assuming public office places at risk all one cares about most in life. A reputation established over the course of decades can be destroyed in an instant. And if later found blameless, an official can only ask—as did an accused and later exonerated member of President Reagan's cabinet—"Where do I go to get my reputation back?" In Washington, in today's polarized environment, it is no longer enough to defeat someone; personal ruin too often becomes the goal. The situation has become so serious that leaders in both parties worry about how to attract high quality people to government service.

And yet they serve.

Nor should we forget the contributions and the families of those who serve. The spouses of our men and women in uniform who remain behind and keep the home functioning by getting the children to school and to little league. Who must shoulder unbearable burdens when a husband or wife is killed or badly wounded. And think about the children of those in uniform who must deal with the loss of a parent or, far more commonly, must often move away from friends and familiar schools, who must deal with the absence of a deployed mom or dad not just on special occasions, but every day.

And while the families of our servicemen and women pay the highest price, the families of others in public service often must deal with the challenges imposed by absences, long hours, and the strains of service. As her husband took on one great responsibility after another, Barbara Bush symbolizes the spouse who kept her family together even while making her own remarkable contributions to society. My wife Becky is another role model in this regard, responding time and again when I have been asked to serve both here and in Washington. She has always said, "We have to do what you have to do." The families of those in public service all "do what they have to do" and pay their own high price.

And yet they too serve.

Why?

Each one in public service has his or her own story—and motives. But I believe, if you scratch deeply enough, you will find that those who serve—no matter how outwardly tough or jaded or egotistical—are, in their heart of hearts, romantics and idealists. And optimists. We actually believe we can make a difference, that we can make the lives of others better, that we can make a positive difference in the life of the greatest country in the history of the world—in President Lincoln's words, "the last, best, hope of earth."

Public servants are people willing to make sacrifices in the present for the future good, people who believe, to paraphrase Walter Lippmann, that we must plant trees we may never get to sit under. They include those of my generation, who heard President Kennedy challenge us nearly 47 years ago, "Ask not what your country can do for you, but what you can do for your country." And people of a later generation heard this president, George H. W. Bush, affirm that "Public service is a noble calling."

Those who serve I think also feel the call of duty. Theodore Roosevelt wrote: "The trumpet call is the most inspiring of all sounds, because it summons men to spurn ease and self-indulgence and timidity, and bids them forth to the field where they must dare and do and die at need." Whatever range of motives causes our young men and women to volunteer for the armed forces, they all hear the call of the trumpet—the call of duty to our country. And because, in doing their duty, they risk all, they are the most noble of all.

But the trumpet summons all of us.

Texas A&M University does much to inculcate a sense of the importance of public service and duty to one's fellow citizens. No institution besides our service academies has consistently commissioned more officers into the military. Texas A&M for years has been a leading source of CIA officers—and that was even before I got here.

Volunteerism is huge here. In the Big Event, Texas A&M sponsors the largest student-run community service project in the nation. In the March to the Brazos, A&M has the largest college fundraiser in the nation for the March of Dimes. In the first year of the Aggie Relay for Life in 2006, Texas A&M held the largest first-time collegiate American Cancer Society event in the nation. There are countless other service projects and ongoing community service efforts. A&M's incredible response to both Hurricanes Katrina and Rita was unforgettable, thousands of students, faculty, and staff volunteers turning large parts of this campus into evacuee centers, hospitals, and a place of refuge.

Washington Monthly magazine three years ago started an annual ranking of schools based on their contribution to the nation and the public good—with criteria such as the percentage of students from lower-income families, percentage of graduates who enter public service and the military, and so on. This year the magazine ranked Texas A&M first in the nation in serving the public good. No one familiar with this unique American institution could have been surprised. Pleased, yes. Surprised? No.

Both here, and on other campuses around the nation, we see evidence every day that young Americans are as decent, generous, and compassionate as we have ever seen in this country. Millions of students all across the country donate countless hours to volunteer organizations, community service, and public-spirited foundations such as the Points of Light Foundation, created by the Bushes.

But I worry—and I worry greatly—that too many young Americans, so public-minded in campus and community affairs, turn aside when it comes to our political process and to careers in public service.

Seventy percent of eligible voters in this country cast a ballot in the election of 1964. The voting age was then 21. During the year I graduated from college, 1965, the first major American combat units arrived in Vietnam and with them, many 18-, 19-, and 20-year-olds. And In recognition of that disparity, later the voting age was lowered to 18.

Sad to say, that precious franchise, purchased and preserved by the blood of hundreds of thousands of college-age Americans—and younger—from 1776 to now, has not been adequately appreciated or exercised by today's young people. In 2004, with the nation embroiled in two difficult and controversial wars abroad, the voting percentage was only 42 percent for those aged 18 to 24.

Too many young people disdain to participate because they believe it won't matter or because they think the system is corrupt or because they think it a waste of time or beneath them. Winston Churchill called those who feign contempt for public affairs "flaccid sea anemones of virtue who can hardly wobble an antenna in the waters of negativity." They don't talk like that anymore. You must rise above that and participate, or else the decisions that affect your life and the future of our country will be made for you—and without you. So, as we approach the 2008 campaign, get involved, and vote.

But to the students in the audience, also consider something else. Consider devoting at least a part of your life to public service.

I have spoken about some of the challenges of public service. As President Bush and I can attest, they are manageable. But the virtue of public service, the benefit, is found not in the size of your house but in the size of your heart. It has always been so. Listen to the words of Abigail Adams, the wife of our second president. She wrote him: "You cannot be, I know, nor do I wish to see you, an inactive spectator ... We have too many high-sounding words and too few actions that correspond with them." Or her letter to her son John Quincy Adams: "These are the times in which a genius would wish to live. It is not in the still calm of life, or the repose of a pacific station, that great characters are formed. The habits of a vigorous mind are found in contending with difficulties. Great necessities call out great virtues."

We live in a time of "great necessities." Our country faces many challenges at home and abroad. It is precisely during these times that America needs its best and brightest, from all walks of life, to step forward and commit to public service. If, in the 21st century, America is to be a force for good in the world—for freedom, social justice, the rule of law, and the inherent value of each person; if America is to be a beacon for all who are oppressed; if America is to exercise global leadership consistent with our better angels, then the most able and idealistic of today's young people must step forward and accept the burden and the duty of public service. You will also find joy and fulfillment—and the ultimate satisfaction that you made a difference.

I quoted earlier from Abigail Adams. I will close with a quote from a letter John Adams sent to their son, Thomas Boylston Adams. He wrote: "Public business, my son, must always be done by somebody. It will be done by somebody or another. If wise men decline it, others will not; if honest men refuse it, others will not."

Will the wise and honest among you come help the American people?

Thank you.

# Sophia University

*Tokyo, Japan, Friday, November 09, 2007*

Good morning. Thank you, ino-sensei, for your warm introduction and this opportunity to be part of Sophia University's lecture series.

As a historian and former university president, I am particularly gratified to be here with you in this center of learning—where ideas are valued, open debate is encouraged, and scholars may pursue knowledge. In fact, since becoming U.S. Secretary of Defense 11 months ago, I have spoken to students at four different universities, including three graduation ceremonies. I have reached out to students because I believe in challenging them—our future leaders—to reach their full potential as responsible citizens of their country and of the world.

I was particularly impressed that Sophia University has more than 500 international students enrolled from more than 50 countries, as well as partnerships with 125 colleges and universities overseas.

This university, in this country, at this time, is a particularly appropriate setting to discuss the security challenges we face together in Asia—challenges that require vibrant and growing partnerships amongst nations of shared values and interests.

On September 4, 1951, President Harry Truman spoke at the opening of the San Francisco conference on the peace treaty with Japan. At the time, a bitter war was raging on the Korean Peninsula and the free world was reconciling itself to a long struggle against Communist expansion. In the week prior, the U.S. had signed mutual defense treaties with three other Pacific nations—Australia, New Zealand, and the Philippines. Truman said something then that I think holds as true today, as we think about meeting Asia's challenges together in the 21st Century. He said:

"In the Pacific, as in other parts of the world, social and economic progress is impossible unless there is a shield which protects men from the paralysis of fear. But our great goal, our major purpose, is not just to build bigger and stronger shields. What we want to do is advance ... the great constructive tasks of human progress."

America's commitment to Asia in the decades since has been sustained through multiple administrations of both political parties. The result has been strategic, political and economic stability that has improved the lives of billions of people. And that commitment is no less strong today, regardless of the challenges we face in the Middle East and elsewhere.

Periodically, we see speculation casting doubt on the future of the United States in Asia, sometimes prompted by changes in U.S. political leadership, or the re-positioning of U.S. military forces. In fact, far from neglecting Asia, we are more engaged than ever before. We have forged, reshaped or renewed security partnerships throughout the Pacific Rim.

In recent years, our security arrangements with Japan have evolved from their Cold War orientation—reflecting both the transformed threat environment and Japan's ability and willingness to play a larger role in its own defense. We have been realigning and repositioning forces here, while cooperating in new areas such as missile defense and sharing new roles and missions with the Japanese Self Defense Force.

Earlier this week, I visited the Republic of Korea for a successful annual Security Consultative Meeting. We are working with South Korea to establish a new vision and force posture that goes beyond the current security situation on the Peninsula and meets the future global needs of both nations. We are preparing for a historic transition in 2012, when the Republic of Korea military will take wartime command in the defense of their own country, and U.S. forces will assume a supporting role.

America's pivotal alliance with Australia is based on a long history of standing together against aggression through multiple conflicts over the past Century. A newer and welcome development is that Australia is taking a larger regional and global role—with their leadership role in East Timor and their deployments in recent years to the Middle East and Central Asia.

Our relationship with India—the world's largest democracy—has evolved from an uneasy coexistence during the Cold War to a growing partnership today. Since the 2005 summit between President Bush and Indian Prime Minister Singh, we have expanded our cooperation in a number of areas, including military to military exchanges.

And then there are nations like Mongolia, once part of the Communist bloc, and now a strong partner in the war on terrorism. Mongolia is also making contribution to global security by supporting UN peace operations and the NATO mission in Kosovo.

Beyond strengthening our traditional alliances and forging new ties, the United States is reinforcing its own capacity in the region. We are investing in new capabilities and infrastructure—gains that will be critical to raising the region's overall ability to respond to security challenges, natural disasters, and potential crises.

In an address to the Japan National Press Club last month, Ambassador Schieffer noted that after the Second World War, Asia's security architecture mostly reflected a "hub and spoke" model, with the U.S. as the "hub" and the spokes representing a series of bilateral alliances with other countries that did not necessarily cooperate much with each other. The U.S. alliance system has been the cornerstone of peace and security in Asia for more than a generation. These alliances are enduring and indispensable. But we would like to see more engagement and cooperation among our allies and security partners—more multilateral ties rather than hubs and spokes. The trilateral dialogue between the United States, Japan and Australia is a good example.

The major challenges facing the region—such as North Korea and nuclear proliferation –cannot be overcome by one, or even two countries, no matter how wealthy and powerful. They require multiple nations of shared interests to come together to deal with a number of key challenges—areas where each partner can bring its unique capabilities to bear for the common good.

Terrorism and violent extremism are a threat to the very fabric of international society, and Asia is not immune. This city suffered a poison gas attack in its transit system in 1995. Then there were the Bali bombings and activities by Islamist groups in the Philippines and Indonesia. It is a long distance from New York's World Trade Center to a Tokyo subway station, but the threat posed by radical groups with violent ideologies is the same. The terrorists have learned to exploit the strengths of modern societies—our technology and infrastructure—and in the case of democracies, our freedoms and our openness as well.

During my visit to Singapore on June 1, I challenged regional leaders to play an even bigger role integrating Central Asia into the Asian security architecture. There are several areas—such as infrastructure development, capacity building, and security assistance—where East and South Asia can do more.

As we've learned on more than one occasion, instability or failed states halfway around the world can have serious implications at home. In no region is this truer than in the Middle East—not just with regard to Iraq, but with the behavior and ambitions of Iran and the operations of terrorist and militia groups. It is worth remembering that Japan imports 80% of its oil from the Gulf to power its economy.

The proliferation of dangerous weapons and materials to terrorists and others is another major threat. The United States knows that we cannot block the flow of these weapons on our own, which is why we work with partners to improve physical security, interdict shipments, and employ sanctions when necessary. The Proliferation Security Initiative is the cornerstone of this effort and it is showing results in Asia.

Maintaining a reliable deterrent against attacks from ballistic missiles is a critical objective of our national security strategy. Building this capability and countering this threat is of special significance to the people of Japan, given the direct danger to your homeland posed by North Korea's weapons programs. As a defensive measure, we are working together to create a network that both deters aggression and provides protection in the case of a missile launch.

We also recognize the importance of maintaining free and secure maritime routes and infrastructure. The Container Security Initiative is one element of our strategy. President Bush also approved the 2005 National Strategy for Maritime Security to help prevent piracy and other hostile and illegal acts within the maritime domain.

Calamities such as the 2004 tsunami, the devastating earthquake in Central Asia, and the typhoons that ravage this part of the world demonstrate the need for a regional approach to humanitarian crises. The U.S. is committed to assisting Asian nations in their time of need, and we are working with partners to prepare and fine-tune our collective response before disaster strikes. We are also preparing for other non-traditional security threats that result from infectious diseases such as pandemics, avian influenza, and SARS.

Although there has not been a single major conflict in Asia for more than three decades, the Northeast corner of the Pacific remains one of the last places on earth with the potential for a nuclear confrontation. We are working with China, Russia, Japan, and the Republic of Korea through the Six-Party-Talks to pressure North Korea to forgo their nuclear ambitions. These talks have had a stabilizing effect on the region in the aftermath of the North's missile and nuclear tests of 2006. We now have a mechanism in place to forge cooperation on the long standing problem of North Korea's behavior and nuclear ambitions.

These challenges are taking place in a context being shaped by the rise and re-emergence of China and Russia—two nations at strategic crossroads taking on a more assertive role in world affairs.

I have just come from China, where I met with senior officials who confirmed their desire to cooperate more in order to address common security challenges. I do not see China as a strategic adversary. It is a competitor in some respects and partner in others. While we candidly acknowledge our differences, it is important to strengthen communications and to engage the Chinese on all facets of our relationship to build mutual understanding and confidence.

We continue to raise with China the need for greater transparency and candor into their strategic military motivations, decision making, and key capabilities. In particular, we remain concerned over the pace and scope of

China's military build-up. This concern is responsible and appropriate. A lack of transparency carries the risk of misunderstanding and miscalculation, and naturally prompts others to take actions as a hedge against uncertainty.

Russia is another relationship where we must overcome a measure of distrust and doubt. When Secretary Rice and I visited Moscow last month, we made some progress towards setting up a regular dialogue to deal with a range of issues that have divided us—including Russia's concerns regarding missile defense. Our proposal to pursue a partnership with Russia in this area was real and sincere. We look for Russia to be equally innovative and forthcoming.

I recall during the 1970s that many people discounted the value of holding strategic talks with the former Soviet Union because these meetings often did not lead directly, or immediately, to new arms control breakthroughs. It turned out that maintaining that dialogue helped each side better understand the other's intentions, and laid the groundwork for gains that ultimately brought the Cold War to a close. The situation we face with Russia and China is nothing close to the old superpower conflict, but the lessons from that period with regard to keeping open lines of communication still apply today.

I would like to close with some thoughts on the future direction of our relationship with Japan.

It has been just over 55 years since Japan was invited to take "her rightful place of equality and honor among the free nations of the world." At the time, President Truman expressed his confidence that "the people of Japan...are ready and willing to play their full part in meeting the common menace."

Japan became a stalwart ally and anchor of democracy and prosperity in Asia through some of the most difficult days of the Cold War. Our friendship held fast through the inevitable turbulences along the way– from political turmoil to sometimes nasty trade disputes.

I recall well the build up to the first Gulf War with Iraq in 1991. Japan contributed significant financial support, but no military forces. At the time, Japan was criticized by some for what was called "checkbook diplomacy." Since then Japan has found more direct ways to contribute to international security. We've seen the support Japan has provided to the people of Afghanistan and Iraq as they rebuild nations torn apart by war, dictatorship and sectarian division.

When my colleague, Secretary of State Condoleezza Rice spoke at this university in 2005, she said "Japanese leadership in advancing freedom is good for the Pacific community, and it is good for the world." Japan has the opportunity—and an obligation—to take on a role that reflects its political, economic, and military capacity. That is why the U.S. strongly supports Japan's becoming a permanent member of the United Nations Security Council. And that is why we hope—and expect—Japan will choose to accept more global security responsibilities in the years ahead.

We must continue to ask each other as partners in this alliance: What should Japan and the U.S. do together, and with others, to secure our mutual interests? Do we have the proper capabilities, individually and collectively, to address future challenges and uncertainties? Have we the proper mechanisms and infrastructure to meet our common objectives? These questions underlie the alliance transformation effort we have undertaken over the last few years. But we need to deepen our discussion and more importantly, be prepared to act on our findings and make the investments now that will better prepare us for the future.

As you can see from the range of issues I've mentioned, the security landscape of the present and foreseeable future will be complex. Your generation will face many challenges—but it will also have undreamt of opportunities, perhaps seen yesterday, barely visible today. What will remain constant is the partnership of shared interests and values between our two nations.

A living example of that friendship, as many of you probably know, has been the cherry tree. In 1912, Japan gave the U.S. over 3,000 cherry trees—including 12 different varieties—as a symbol of friendship and goodwill. The wives of the American President and the Japanese Ambassador planted the first two trees along the Tidal Basin in Washington. The remaining trees were planted in a ring just across the Potomac River. What is less well known is that seventy years and two World Wars later, the United States had an opportunity to reciprocate the gesture, when we provided cuttings from our trees to replace those destroyed by a flood in Japan.

Next spring, from my office at the Pentagon, I will be able to see the trees from my window. Every day, people from around the world visit Washington to view thousands of pink and white flowers in bloom. And the first two trees—planted almost a century ago—still stand today, a living symbol of the enduring ties between our countries.

Thank you. I look forward to your comments and questions.

## Military Spouse Career Advancement Initiative

*Pentagon Conference Center, Wednesday, November 14, 2007*

GOOD morning everyone, and thank you all for coming. Today we embark on a landmark program that will open the doors to our military spouses for more fulfilling careers.

We understand how hard it is for families of military men and women who have to be ready to move anywhere in the world every couple of years. Spouses are called on to pack up and relocate the family often at the cost of their own careers.

This makes it difficult to navigate the career licensure and certification requirements that go with most professions. In addition, education is often unaffordable for young families, who must also bear the expenses of child care.

The Department of Defense has long been committed to helping military spouses pursue rewarding careers and removing barriers to employment. We have reached out to employers and helped them recognize the true value of military spouses. In the past couple of years, as David [Chu] mentioned, more than 400 companies, many of whom are represented here today, have committed to hiring military spouses. Even greater potential lies in the collaboration we embark on today.

The Department of Defense could not have done this alone. In collaboration with the Department of Labor, we are investing more than $35 million in the initial phase of this initiative which will take place in eight key states with 18 military installations.

We appreciate the support of everyone who has made this program possible. Specifically, I want to thank our partner, Elaine Chao, who as Secretary of Labor has long been a champion of this effort, and who has made it a point to specifically help our military spouses.

Beginning this January, the Departments of Defense and Labor will launch the test phase of the Military Spouse Career Advancement Initiative. This initiative will help military spouses address education, training, and professional licensing and certification issues needed to have high-growth, portable careers in fields such as technology and health care. Community colleges on military installations are valued participants this endeavor as well. I would like to thank the representatives here today for their support in preparing military spouses for new careers.

We owe it to our brave men and women in uniform to assist their families as they do their job, often thousands of miles from home and their families and under extremely hazardous conditions. When service members find the time to call home or email home, they shouldn't have to worry. They have the right to hear their loved ones honestly say, "We miss you, but we are doing fine."

Thank you, and now I am now pleased to present Secretary Chao.

# Boy Scouts of America "Citizen of the Year" Award

*Washington, DC, Thursday, November 15, 2007*

THANK you, Secretary West. What Togo failed to mention was that he was one of the co-chairs of that group that investigated the problems of the care of our wounded warriors. I thank you for your service.

Antar, you are amazing. You are our future. We admire you and, apropos of the comments that were made, I think you're a great dad model.

Thank you all for coming and inviting me here tonight. It is an honor to receive the "Citizen of the Year" award from the National Capital Area Council. When I say it's a pleasure to be here, I mean it. I don't face many friendly audiences in this city.

Nearly 15 years ago, I re-engaged with scouting after a hiatus of nearly 30 years, when retired Chief of Staff of the Air Force and then-Chairman of the National Capital Area Council board General Larry Welch asked me to serve on your board. I told him I would be moving to the Pacific Northwest within a year or so, and he told me that did not matter. So, I joined. Some years later, I was asked why I had not gotten re-involved in scouting much earlier. I said, because I was busy and no one ever asked me. Let that be a lesson. Never hesitate to ask. Anyway, in the summer of 1994 I headed for the Pacific Northwest and I thought, retirement. Somehow, it just didn't work out. So tonight, as I have done before some 130 Scout Councils including this one, I would like to speak briefly about the impact scouting has had on me.

First, scouting was really the only preparation early on that I ever had for becoming a leader. I attended the National Junior Leader Training program in Philmont when I was a teenager—a program that taught me how to deal with people, how to set goals, and then go about achieving them. I actually never saw the need for another management course. All would only have been wordier, more expensive versions of what I learned at age 15 at JLT at Philmont.

Second, scouting taught me the importance of character. I've never—in the nearly 50 years since I was a Boy Scout—found a better expression of the way people ought to live their lives outside of the Bible than the scout oath and law.

Third, scouting gave me a sense of personal responsibility, not only to other people, but also to the world around me—nature, community, and the country. The same ethic I learned as a Boy Scout to leave a campsite better than I find it, I have tried to apply to life in general—to leave every situation better than when I found it. This is not as easy as it sounds, in government, in a university, or in business. You have to feel this deep inside.

In this era of YouTube, violent video games, and celebrities lining up for re-hab—a time when my press conferences seem strangely and regularly pre-empted by late-breaking important news of the doings of Paris Hilton or Britney Spears—in such an age, scouting can sometimes seem like a quaint relic of the past. But I would argue that scouting is more. It is more important, more relevant today than ever before.

Where will our young men learn leadership?
Where will they learn character?
Where will they learn personal responsibility?
Where will they learn a spirit of adventure, of taking sensible risks? Of pushing themselves beyond what they thought they could do?

I know that I, personally, learned these lessons first in Boy Scouts, and have used them every step of the journey that led to where I am today.

As a teenager, I was a good but not a great student. I certainly was not a great athlete. Yet, I differentiated myself by earning my Eagle Scout badge, an achievement which required, and demonstrated, as for all those who earn it, persistence, successful goal-setting, and self-discipline. That early achievement gave me the confidence to tackle the increasingly complex challenges that I would face later in life.

For those of you who are scout leaders and wonder if you have had an impact, I can assure you that almost five decades later I can remember all the names and faces of all of my scout leaders—although I must say that's a bigger and bigger feat as the years go on.

A particularly valuable lesson taught to me and my fellow troop members one cold Kansas winter was when our Scout Master, Forest Becken, taught us how to cook on a fire of dried cow chips. It imparted a very distinctive taste to already nearly inedible food … believe me when I say it is a taste I have recognized more than a few times serving here in Washington, D.C.

Leadership skills, and the understanding of the critical significance of character and personal responsibility are three of the most important gifts I ever received, and I received them from the Boy Scouts of America.

I see these attributes displayed by the brave men and women of our armed forces who serve and sacrifice every day in battle against an unrelenting enemy determined to do our country harm.

While we are a nation at war, we are also in the middle of another war, this one inside America. It is a battle for the souls of our boys and young men. I firmly believe every boy who joins the Scouts is a boy on the right track.

While I believe that we are facing a world of unprecedented threats, I also believe we face a world full of unprecedented hope and opportunity. Scouting provides the kind of optimistic, confident and skilled young leaders of integrity who will ensure that we fulfill the hope and seize the opportunity.

I'll close with a story that conveys this sense of hope about America's youth and America's future. It was December 1985, and I was at CIA during what was a particularly bleak period. There were a lot of terrorist attacks.

That December, CIA received a letter from a group of Boy Scouts, and they offered to help us fight the terrorists—but only on one condition: that we didn't tell their mothers they had written to us.

As one senior CIA officer at a staff meeting observed, at least they had a well-developed sense of security.

I'll wager at least some of those boys—now men—are, in fact, as I speak, in an American military uniform somewhere fighting terrorists. Their moms know it this time, and we pray for them all.

I'm honored by your "Citizen of the Year" award, and will do my best to live up to your expectations for a Boy Scout from Kansas.

It will occupy the same place of honor as my Silver Buffalo Award. These objects will be proudly displayed. But make no mistake, they both pale in comparison to the real reward—the lessons the Boy Scouts taught me in the days of my youth. I only hope I can live up to my very first oath—the Scout Oath—made more than half a century ago, and continue to be worthy of this award. Thank you.

## Base Community Council

*Whiteman Air Force Base, Missouri, Tuesday, November 20, 2007*

THANK you for that warm welcome, Congressman Skelton, and special thanks to Whiteman's Base Community Council for your invitation.

Let me say a word or two about my friend, Ike Skelton, probably nothing you don't already know. A representative of Missouri's fourth district since 1977, Ike Skelton has always defended the interests of our men and women in uniform and is a life-long champion of American military power. A strong advocate for Fort Leonard Wood, the Missouri National Guard Training Center, and of course Whiteman, the Congressman even has a picture of the B-2 on his homepage!

As Chairman of the House Armed Services Committee, he is the leader on a number of issues vital to our nation's security, among them nuclear non-proliferation and reconstituting the Armed Forces. He has visited the Persian Gulf and met with leaders of Iraq and Afghanistan to assess the situation firsthand. The House Armed Services Committee has long been a source of support in meeting our nation's defense goals. Following my Senate confirmation as the new Secretary of Defense, the first committee that called me to testify was the House Armed Services Committee. The subject was "The Way Forward in Iraq"—not an easy pitch or an easy audience. I can assure you that your Congressman's questions are tough; they are pointed—but they are always fair.

In a world of partisan politics, Ike has always placed the welfare of our service members at the forefront of his agenda. He has sat bed-side at Walter Reed and comforted those in the amputee wards. Last year, he visited troops in Iraq during the Thanksgiving holiday. All told, Ike Skelton is a friend of everyone in uniform. I deeply appreciate his efforts to safeguard this great nation and his enduring concern for our service members. Thank you.

Well, it is great to be here in the mid-west—it feels like coming home. As Congressman Skelton mentioned, I grew up just next door and got one degree from Indiana University in Bloomington, Indiana. With reference to Ike's first comment, I will remain silent about Saturday's game. I will tell you that my brother, my sister-in-law, and my niece all have degrees from Kansas

[State] University—none of them from KU—all K-State. The only reaction I have to Saturday's game is, as the former President of a university with a record so far this season of 6-5 and who is going to play the University of Texas on Friday, I'm simply envious of both team's records so far this year.

There is a feeling of pride in America's heartland that I have always admired and appreciated. You see examples of integrity, loyalty, and patriotism all around you. I learned firsthand about the importance of close neighbors and tight-knit communities like this one.

Being here and by the way, I should tell you my father was born and grew up in Missouri. In the 1920s, [he] sold typewriters, driving model Ts through the Ozarks. He had some interesting stories to tell. Being here in Missouri is coming full circle from Air Force second lieutenant to Secretary of Defense. I was commissioned on January 4, 1967. I married my wife, Becky, in Seattle on January 7, 1967 and a few days later reported for duty here at Whiteman. I would share with you that the largest fixed wing aircraft we had here at the time was a Cessna. I was assigned to the base plans and intelligence office. One of my duties was to brief missile crews on international political and military developments ... I would tell you their lack of interest was awesome.

Because of my academic background and moderate Russian language skills, I frequently briefed high-ranking officers on our wing's Minuteman targets in the Soviet Union. I recall one briefing in particular. I was explaining our target set to an Air Force Lieutenant General, the Commander of 8th Air Force at Westover—whom I would characterize as a cigar-chomping, foul-mouthed Curtis Lemay wannabe. When I told him that 120 of our 150 missiles were aimed at Soviet ICBMs, he blew up and, with many expletives I will delete, said it was an outrage that we would be hitting only empty silos. He wanted to kill Russians. He demanded that I, Second Lieutenant Gates, rewrite the nuclear targeting plan. That reminded me there was a four-star general in Omaha who felt he had some ownership of that problem.

That does remind me of another story about targeting. One day here at Whiteman, we told there was a problem with the war plans. SAC Headquarters in Omaha needed to change the launch sequencing for all the missiles immediately. So, we worked all night to fix the strike control documents, then—wrestling with large, unwieldy sheets of laminating material like working with fly paper. We laminated the documents and checklists. The next morning, we received a call from a major in one of the launch control capsules. He correctly guessed the kind of pizza we had eaten while working during the night ... there seemed to be a piece of pepperoni laminated under a strike control sheet.

I still remember that, when I arrived at Whiteman, many Missouri farmers were unhappy about the missiles because two acres of their land had been taken for each of the silos. One story I heard here was that, as a way to smooth over relations, the Air Force flew some of the farmers to the ballistic missile staff officer course at Vandenberg Air Force Base in California. There the farmers saw a video showing a silo door, about 80 tons of steel and concrete, being blown off the top of a silo. One of the farmers was convinced the silo door being blown aside like this would kill some of his cows. The Air Force tried to persuade him that if that silo door blew he had bigger problems to worry about than his cows, but to no avail. And so, to calm his fears, base personnel anchored two telephone poles outside the silo fence on his property—in theory—to stop the silo cap from hurting his cows.

On another occasion, we were told, one of the base helicopters, which were used to ferry crews and secret documents to the launch control capsules, was forced to land in an adjacent field because of high winds. A still resentful, no-nonsense Missouri farmer drove his tractor over, then chained and padlocked a strut of the helicopter to his tractor, and demanded $1,000 in rent for use of his field as a helo pad.

Relations between the base and community presumably have improved so much since then. I'm delighted to hear about the ways that you, members of the Base Community Council, have embraced our service men and women. With representatives from over 20 surrounding cities and towns—both large and small—the BCC has:

- Helped fund nearly two dozen squadron Christmas parties on base each year;
- Rented caps and gowns for Airmen graduating from the Community College of the Air Force; and
- Volunteered at air shows and other base-wide community events.

I travel to Afghanistan and Iraq fairly regularly and I am routinely asked by troops whether the people back home support them. I tell them 'yes' because of organizations like yours. To everyone here from the BCC, please accept my heart-felt thanks for all the wonderful ways you have volunteered to help this military community.

We are the most powerful military in the world today because of the service members who have volunteered to answer the nation's call. The men and women stationed here fly or support B-2s, A-10s, T-38s and Apaches—awesome instruments in the arsenal of freedom. When Harry Truman said, "carry the battle to them; don't let them bring it to you," he did not envision 44-hour round trip missions by B-2s made from this base to targets on the other side of the globe.

Just as Whiteman provided strategic depth and deterrence during the Cold War, so too do its platforms and its people safeguard America today. While consolidating gains in Afghanistan is a priority and Iraq remains at the forefront in the war on terror, the strategic deterrence and long-range, precision capabilities offered here—our only operational B-2 base—are critical to protecting America's other national interests around the world.

I will conclude with one final story that illustrates what else has changed. In the mid-1990s, I was giving some speeches in Hanoi. Following one, Vietnam's Minister of Defense approached me and asked if I had fought in Vietnam during the war. I replied, "No, I was stationed at an intercontinental ballistic missile base in Missouri. I was targeting your patron, the Soviet Union, in case they got out of hand." The Minister's face froze a few seconds, then he slowly broke a smile and said, "That's good, that's good!"

Congressman Skelton and members of the BCC, again, thank you for showing our troops how much you care. Now I'd be happy to answer your questions. Thank you.

## Greater Killeen Chamber of Commerce

*Killeen, TX, Monday, November 26, 2007*

THANK you for that kind introduction, Pat, and I also want to thank the Greater Killeen Chamber of Commerce and the Heart of Texas Defense Alliance for inviting me to this event.

It is great to be in Killeen, and to be back in Central Texas. We are not far from College Station, where I spent four and a half wonderful years as president of Texas A&M. And it is clear that there are some Aggies with us tonight. Of course, visiting Texas in November is another reminder that out here, football is practically a form of organized religion. The first fall when I was at Texas A&M, and we changed the football coach, I told the press that I had overthrown the governments of medium-sized countries with less controversy. And Chancellor McKinney, I'm glad I'm not there now.

Of course the Texas Hill country has it all over Washington. A place where, as Senator Alan Simpson used to say, "Those who travel the high road of humility encounter little heavy traffic." A place where there are so many lost in thought because it's such unfamiliar territory. Where people say "I'll double-cross that bridge when I get to it." The only place in the world where you can see a prominent person walking down lover's lane holding his own hand.

This is my first visit to Fort Hood as Secretary of Defense. And it is an honor. Few Army posts have given as much to the defense of this nation as Fort Hood, home to some of the biggest and most well-known units of the United States military. And I would have to say there are a couple of legends here tonight. First and foremost, General Bob Shoemaker. And I would add—and he would be very embarrassed by this—also Lieutenant General Pete Chiarelli. Who cannot be awed by men with such records of service, and sacrifice, and success. But General Shoemaker, you may need, as you may have hinted to me earlier, to recruit a better class of football players for the high school named for you.

Deployments from this post—of the First Cavalry Division, the Fourth Infantry Division, and other elements of the "Third Corps"– have been constant. Come this spring, it will be five solid years that at least one of these two divisions has been deployed in support of the Global War on Terror. They have been in the thick of the battle against the jihadists—fighting in some of the toughest campaigns and doing some of the most difficult work to bring stability and reconstruction to the fledgling democracies in Afghanistan and Iraq.

The troops of the Fourth ID were in Tikrit when a dictator was fished out of a hole in the ground. And the men and women of the "Iron Horse Division" were there when Zarqawi—one of the world's most vicious and dangerous terrorists—met his end.

"The First Team," as it is called, quelled the uprising in Baghdad's Sadr City in 2004, and stayed on to bring a measure of security and improved conditions to the people there. The First Cav was led at that time, as you know, by General Chiarelli.

The troops from these and other Fort Hood units watched Iraqis vote for the first time in real elections, adopt a constitution, seat a democratically elected prime minister, and form army and police forces that have increasingly taken on the task of securing their country. Acts of ethnic and sectarian violence have diminished in recent months, in part due to the work of tens of thousands of soldiers from this post and across the nation.

Tonight I'm here for a special reason, and that is simply to acknowledge what it is that makes all of those accomplishments possible: the support that Fort Hood soldiers receive from their families, and that both soldiers and the families receive from the people of this community.

America owes a great deal to those who have been called "the power behind the power"—the spouses, children, parents, grandparents, brothers and sisters of our men and women in uniform. They, too, make a significant contribution—and pay a price—in the cause of protecting the United States and its allies. They are the ones who remain back home, wrestling with the challenges of day-to-day life that come up in families and households that have a very important member missing. As President Bush said of military spouses, their service to the country "begins as soon as they say two words: 'I do.'" The multiple overseas deployments out of Fort Hood have lengthened family separations and put an even heavier burden on Fort Hood spouses and other kin.

And yet they have borne that burden with grace and patience and an amazing ability to organize and rely upon one another. We see this in so many ways, not least in the "care teams" made up of family members from the units that go to those who have been notified of the death or injury of a loved one. The care team members are there to help them grieve, offering a kind word, a cooked meal, a shoulder to cry on. Army families take care of their own and Fort Hood families are no exception. They are strong. They endure. They are bound together by their shared experiences, by sacrifice, and by the pride they rightly feel in the noble work their soldiers do. They are an example to us all.

But military family members are not, and could not be, completely self-sufficient. That's where many of you come in. Fort Hood has grown by over 10,000 soldiers since 2003, and has added more families and more children. But instead of calling for the federal government to respond, the citizens of Central Texas stepped up, on their own, to welcome the newcomers.

Acts of kindness that make headlines in other parts of the country are engrained in the very fabric of your communities. The church dinners, the social events, cutting the lawn for a family whose soldier is deployed—hundreds of events, large ones and small, every day of the week, for five years. You do these things without being asked and without asking anything in return.

Community members faithfully attend ceremonies that are held on the post for changes of command, the awarding of the Purple Heart, retirements, and other milestones.

They are also making weekly trips to the Brooke Army Medical Center to visit wounded Fort Hood soldiers and present them with caps, and t-shirts, and bathrobes. Those items don't come free, you know; and I understand the cheerleading squad of Harker Heights High School raised money to help purchase them.

Then there are the more than 490 Fort Hood units that have been "adopted" by businesses, individuals, and civic clubs of Central Texas through the Association of the U.S. Army. The local AUSA chapter and Sprint have joined forces to give scholarships to soldiers and their families.

The Chambers of Commerce of Greater Killeen, Belton, Temple, and many other towns and cities, also give scholarships, as well as initiating awards such as Soldier-of-the-Quarter.

There is also the "Living in the New Normal" initiative of the Military Child Education Coalition, which started right here as a coalition between Fort Hood and the Killeen Independent School District. It grew from that small seed, and is now an extensive non-profit organization that helps military kids across the United States and abroad, by making counseling and other services available to these children to ease their frequent transitions to new schools.

Through the teamwork of Central Texas communities, Fort Hood, and Dell, 50 spouses of Fort Hood soldiers became virtual customer care agents for Dell under the Army Spouse Employment Partnership program.

Every one of these forms of assistance matters. This is a tough time for our troops and their families—and I can assure you that your aid and your assistance does not go unnoticed.

Thinking back in time, to the late 60s and 1970s, we remember that civilian appreciation for the United States military has not always been there. As one who was in the military at the time, I cannot express to you how heartwarming it is to see civilians united today in their admiration of our men and women who have volunteered to serve at such a challenging time for our country. You see it in airports all over the country, where service men and women are met with standing ovations by passengers in the terminal. There are free meals and rounds of drinks, at least only for soldiers over 21. And, above all, simple thank yous. The appreciation is real, it is sincere, and it bridges any political divide.

No one can tell what lies ahead for our nation in the conflicts in which we are engaged. But we do know that great challenges remain.

An immediate challenge the Defense Department faces is the ongoing delay in the Congress over the war funding bill. The facts are simple. Without these funds, Army operations and maintenance accounts will be exhausted by mid-February, and similar Marine Corps accounts about a month later. We cannot wait until mid-February to figure out how to deal with the consequences of these accounts running dry. For example, under certain of our contracts, civilian employees must be notified sixty days in advance of a furlough, and that means mid-December. We're not trying to scare anyone, or play politics. That's not the way I do business. But I am responsible for prudent management and planning. And that means prior planning just in case we don't get this funding in a bill that this President will sign. The Defense Department is like the world's biggest supertanker. It cannot turn on a dime, and I cannot steer it like a skiff. I don't want to create anxiety among our employees. But we have to plan, and we have to prepare.

Turning to a brighter subject, one result of the improved Iraqi security situation is the beginning of a lowering of U.S. troop levels. A brigade of the 1st Cav is in the process of coming home, and its battle space is being assumed by another unit in country. The members of that brigade, and of every other unit throughout the U.S. armed forces, have been giving this effort everything they've got. And they've gained something in return: They know that they are defending our country and shaping the course of history.

Let me point out, as well, that more help is on the way: with bipartisan support in the Congress, a bigger Army, more funds to mitigate the stress of six years at war, and new programs and commitments to make the Army's covenant with families a reality.

We want every family with a loved one overseas right now to be reunited with that soldier and we look forward to the day when it will be possible. Until then, Fort Hood soldiers and their families will be doing what they have always done: their duty. I express my deepest admiration for, gratitude to, and pride in the soldiers and the families of Fort Hood, and the generous community that stands behind them.

Thank you.

# Landon Lecture

*Manhattan, Kansas, Monday, November 26, 2007*

THANK you, Jon for that kind introduction, and for inviting me to speak here today.

Congresswoman Boyda, Speaker Neufeld, thank you for being here. It is also good to see General Durbin and soldiers with us from Fort Riley and Leavenworth.

I'd like to extend a special thanks to the ROTC cadets in the audience. Your willingness to serve in this time of peril is a testament not only to yourselves, but to a new generation of leaders who will face great challenges in the coming years.

It is both an honor and a pleasure to be part of the Landon Lecture series—a forum that for more than four decades has hosted some of America's leading intellectuals and statesmen. Considering that fact, I at first wondered if the invitation was in fact meant for Bill Gates.

It is a pleasure to get out of Washington, D.C., for a little while. I left Washington in 1994, and I was certain, and very happy, that it was the last time I would ever live there. But history, and current events, have a way of exacting revenge on those who say "never." I've now been back in the District of Columbia for close to a year, which reminds me of an old saying: For the first six months you're in Washington, you wonder how the hell you ever got there. For the next six months, you wonder how the hell the rest of them ever got there.

As I look down at my remarks and the material I want to cover this afternoon, I am reminded of the time George Bernard Shaw told a speaker he had 15 minutes to speak. The speaker replied, "15 minutes? How can I tell them all I know in 15 minutes?" Shaw responded, "I advise you to speak very slowly." I want to warn you in advance that my remarks are more than 15 minutes.

Dr. Wefald has highlighted my K-State bona fides. I would just comment that my mother who is 94 attended my swearing-in ceremony in Washington. That night Conan O'Brian remarked on the fact that I had announced that my 94 year-old mother was there and then he said, "she came up to me and said...'now go beat the hell out of the Kaiser.'"

It is good to be back in Kansas, where my family has lived for more than a century.

I believe Kansas imparts to its children three characteristics that have been a source of strength for me over the years: a rejection of cynicism and an enduring optimism and idealism.

Looking around the world today, optimism and idealism would not seem to have much of a place at the table. There is no shortage of anxiety about where our nation is headed and what its role will be in the 21st century.

But I can remember clearly other times in my life when such dark sentiments were prevalent. In 1957, when I was at Wichita High School East, the Soviet Union launched Sputnik, and Americans feared being left behind in the space race and, even more worrisome, the missile race.

IN 1968, THE FIRST FULL YEAR I LIVED IN WASHINGTON, WAS THE SAME YEAR AS THE TET OFFENSIVE IN VIETNAM, WHERE AMERICAN TROOP LEVELS AND CASUALTIES WERE AT THEIR HEIGHT. ACROSS THE NATION, PROTESTS AND VIOLENCE OVER VIETNAM ENGULFED AMERICA'S CITIES AND CAMPUSES. ON MY SECOND DAY OF WORK AS A CIA ANALYST, THE SOVIET UNION INVADED CZECHOSLOVAKIA. AND THEN CAME THE 1970S—WHEN IT SEEMED THAT EVERYTHING THAT COULD GO WRONG FOR AMERICA DID.

Yet, through it all, there was another storyline, one not then apparent. During those same years, the elements were in place and forces were at work that would eventually lead to victory in the Cold War—a victory achieved not by any one party or any single president, but by a series of decisions, choices, and institutions that bridged decades, generations, and administrations. From:

- The first brave stand taken by Harry Truman with the doctrine of containment; to
- The Helsinki Accords under Gerald Ford; to
- The elevation of human rights under Jimmy Carter; to
- The muscular words and deeds of Ronald Reagan; and to
- The masterful endgame diplomacy of George H. W. Bush.

All contributed to bring an Evil Empire crashing down not with a bang but with a whimper. And virtually without a shot being fired.

In this great effort, institutions, as much as people and policies, played a key role. Many of those key organizations were created 60 years ago this year with the National Security Act of 1947—a single act of legislation which established the Central Intelligence Agency, the National Security Council, the United States Air Force, and what is now known as the Department of Defense. I mention all this because that legislation and those instruments of national power were designed at the dawn of a new era in international relations for the United States—an era dominated by the Cold War.

The end of the Cold War, and the attacks of September 11, marked the dawn of another new era in international relations—an era whose challenges may be unprecedented in complexity and scope.

In important respects, the great struggles of the 20th century—World War I and World War II and the Cold War—covered over conflicts that had boiled and seethed and provoked war and instability for centuries before 1914: ethnic strife, religious wars, independence movements, and, especially in the last quarter of the 19th century, terrorism. The First World War was, itself, sparked by a terrorist assassination motivated by an ethnic group seeking independence.

These old hatreds and conflicts were buried alive during and after the Great War. But, like monsters in science fiction, they have returned from the grave to threaten peace and stability around the world. Think of the slaughter in the Balkans as Yugoslavia broke up in the 1990s. Even now, we worry about the implications of Kosovo's independence in the next few weeks for Europe, Serbia, and Russia. That cast of characters sounds disturbingly familiar even at a century's remove.

The long years of religious warfare in Europe between Protestant and Catholic Christians find eerie contemporary echoes in the growing Sunni versus Shia contest for Islamic hearts and minds in the Middle East, the Persian Gulf, and Southwest Asia.

We also have forgotten that between Abraham Lincoln and John F. Kennedy, two American presidents and one presidential candidate were assassinated or attacked by terrorists—as were various tsars, empresses, princes, and, on a fateful day in June 1914, an archduke. Other acts of terrorism were commonplace in Europe and Russia in the latter part of the 19th century.

So, history was not dead at the end of theCold War. Instead, it was reawakening with a vengeance. And, the revived monsters of the past have returned far stronger and more dangerous than before because of modern technology—both for communication and for destruction—and to a world that is far more closely connected and interdependent than the world of 1914.

Unfortunately, the dangers and challenges of old have been joined by new forces of instability and conflict, among them:

- A new and more malignant form of global terrorism rooted in extremist and violent jihadism;
- New manifestations of ethnic, tribal, and sectarian conflict all over the world;
- The proliferation of weapons of mass destruction;
- Failed and failing states;

- States enriched with oil profits and discontented with the current international order; and
- Centrifugal forces in other countries that threaten national unity, stability, and internal peace—but also with implications for regional and global security.

Worldwide, there are authoritarian regimes facing increasingly restive populations seeking political freedom as well as a better standard of living. And finally, we see both emergent and resurgent great powers whose future path is still unclear.

One of my favorite lines is that experience is the ability to recognize a mistake when you make it again. Four times in the last century the United States has come to the end of a war, concluded that the nature of man and the world had changed for the better, and turned inward, unilaterally disarming and dismantling institutions important to our national security—in the process, giving ourselves a so-called "peace" dividend. Four times we chose to forget history.

Isaac Barrow once wrote, "How like a paradise the world would be, flourishing in joy and rest, if men would cheerfully conspire in affection and helpfully contribute to each other's content: and how like a savage wilderness now it is, when, like wild beasts, they vex and persecute, worry and devour each other." He wrote that in the late 1600s. Or, listen to the words of Sir William Stephenson, author of A Man Called Intrepid and a key figure in the Allied victory in World War II. He wrote, "Perhaps a day will dawn when tyrants can no longer threaten the liberty of any people, when the function of all nations, however varied their ideologies, will be to enhance life, not to control it. If such a condition is possible it is in a future too far distant to foresee."

After September 11th, the United States re-armed and again strengthened our intelligence capabilities. It will be critically important to sustain those capabilities in the future—it will be important not to make the same mistake a fifth time.

But, my message today is not about the defense budget or military power. My message is that if we are to meet the myriad challenges around the world in the coming decades, this country must strengthen other important elements of national power both institutionally and financially, and create the capability to integrate and apply all of the elements of national power to problems and challenges abroad. In short, based on my experience serving seven presidents, as a former Director of CIA and now as Secretary of Defense, I am here to make the case for strengthening our capacity to use "soft" power and for better integrating it with "hard" power.

One of the most important lessons of the wars in Iraq and Afghanistan is that military success is not sufficient to win: economic development, institution-building and the rule of law, promoting internal reconciliation, good governance, providing basic services to the people, training and equipping indigenous military and police forces, strategic communications, and more—these, along with security, are essential ingredients for long-term success. Accomplishing all of these tasks will be necessary to meet the diverse challenges I have described.

So, we must urgently devote time, energy, and thought to how we better organize ourselves to meet the international challenges of the present and the future—the world you students will inherit and lead.

I spoke a few moments ago about the landmark National Security Act of 1947 and the institutions created to fight the Cold War. In light of the challenges I have just discussed, I would like to pose a question: if there were to be a "National Security Act of 2007," looking beyond the crush of day-to-day headlines, what problems must it address, what capabilities ought it create or improve, where should it lead our government as we look to the future? What new institutions do we need for this post Cold War world?

As an old Cold Warrior with a doctorate in history, I hope you'll indulge me as I take a step back in time. Because context is important, as many of the goals, successes, and failures from the Cold War are instructive in considering how we might better focus energies and resources—especially the ways in which our nation can influence the rest of the world to help protect our security and advance our interests and values.

What we consider today to be the key elements and instruments of national power trace their beginnings to the mid-1940s, to a time when the government was digesting lessons learned during World War II. Looking back, people often forget that the war effort—though victorious—was hampered and hamstrung by divisions and dysfunction. Franklin Roosevelt quipped that trying to get the Navy, which was its own cabinet department at the time, to change was akin to hitting a featherbed: "You punch it with your right and you punch it with your left until you are finally exhausted," he said, "and then you find the damn bed just as it was before." And Harry Truman noted that if the Navy and Army had fought as hard against the Germans as they had fought against each other, the war would have been over much sooner.

This record drove the thinking behind the 1947 National Security Act, which attempted to fix the systemic failures that had plagued the government and military during World War II—while reviving capabilities and setting the stage for a struggle against the Soviet Union that seemed more inevitable each passing day.

The 1947 Act acknowledged that we had been over-zealous in our desire to shut down capabilities that had been so valuable during the war—most of America's intelligence and information assets disappeared as soon as the guns fell silent. The Office of Strategic Services—the war intelligence agency—was axed, as was the Office of War Information. In 1947, OSS returned as CIA, but it would be years before we restored our communications capabilities by creating the United States Information Agency.

There is in many quarters the tendency to see that period as the pinnacle of wise governance and savvy statecraft. As I wrote a number of years ago, "Looking back, it all seem[ed] so easy, so painless, so inevitable." It was anything but.

Consider that the creation of the National Military Establishment in 1947—the Department of Defense—was meant to improve unity among the military services. It didn't. A mere two years later the Congress had to pass another law because the Joint Chiefs of Staff were anything but joint. And there was no chairman to referee the constant disputes.

At the beginning, the Secretary of Defense had little real power—despite an exalted title. The law forbad him from having a military staff and limited him to three civilian assistants. These days, it takes that many to sort my mail.

Throughout the long, twilight struggle of the Cold War, the various parts of the government did not communicate or coordinate very well with each other. There were military, intelligence, and diplomatic failures in Korea, Vietnam, Iran, Grenada, and many other places. Getting the military services to work together was a recurring battle that had to be addressed time and again, and was only really resolved by legislation in 1986.

But despite the problems, we realized, as we had during World War II, that the nature of the conflict required us to develop key capabilities and institutions—many of them non-military. The Marshall Plan and later the United States Agency for International Development acknowledged the role of economics in the world; the CIA the role of intelligence; and the United States Information Agency the fact that the conflict would play out as much in hearts and minds as it would on any battlefield.

The key, over time, was to devote the necessary resources—people and money—and get enough things right while maintaining the ability to recover from mistakes along the way. Ultimately, our endurance paid off and the Soviet Union crumbled, and the decades-long Cold War ended.

However, during the 1990s, with the complicity of both the Congress and the White House, key instruments of America's national power once again were allowed to wither or were abandoned. Most people are familiar with cutbacks in the military and intelligence—including sweeping reductions in manpower, nearly 40 percent in the active army, 30 percent in CIA's clandestine service and spies.

What is not as well-known, and arguably even more shortsighted, was the gutting of America's ability to engage, assist, and communicate with other parts of the world—the "soft power," which had been so important throughout the Cold War. The State Department froze the hiring of new Foreign Service officers for a period of time. The United States Agency for International Development saw deep staff cuts—its permanent staff dropping from a high of 15,000 during Vietnam to about 3,000 in the 1990s. And the U.S. Information Agency was abolished as an independent entity, split into pieces, and many of its capabilities folded into a small corner of the State Department.

Even as we throttled back, the world became more unstable, turbulent, and unpredictable than during the Cold War years. And then came the attacks of September 11, 2001, one of those rare life-changing dates, a shock so great that it appears to have shifted the tectonic plates of history. That day abruptly ended the false peace of the 1990s as well as our "holiday from history."

As is often the case after such momentous events, it has taken some years for the contour lines of the international arena to become clear. What we do know is that the threats and challenges we will face abroad in the first decades of the 21st century will extend well beyond the traditional domain of any single government agency.

The real challenges we have seen emerge since the end of the Cold War—from Somalia to the Balkans, Iraq, Afghanistan, and elsewhere—make clear we in defense need to change our priorities to be better able to deal with the prevalence of what is called "asymmetric warfare." As I told an Army gathering last month, it is hard to conceive of any country challenging the United States directly in conventional military terms—at least for some years to come. Indeed, history shows us that smaller, irregular forces—insurgents, guerrillas, terrorists—have for centuries found ways to harass and frustrate larger, regular armies and sow chaos.

We can expect that asymmetric warfare will be the mainstay of the contemporary battlefield for some time. These conflicts will be fundamentally political in nature, and require the application of all elements of national power. Success will be less a matter of imposing one's will and more a function of shaping behavior—of friends, adversaries, and most importantly, the people in between.

Arguably the most important military component in the War on Terror is not the fighting we do ourselves, but how well we enable and empower our partners to defend and govern themselves. The standing up and mentoring of indigenous army and police—once the province of Special Forces—is now a key mission for the military as a whole.

But these new threats also require our government to operate as a whole differently—to act with unity, agility, and creativity. And they will require considerably more resources devoted to America's non-military instruments of power.

So, what are the capabilities, institutions, and priorities our nation must collectively address—through both the executive and legislative branches, as well as the people they serve?

I would like to start with an observation. Governments of all stripes seem to have great difficulty summoning the will—and the resources—to deal even with threats that are obvious and likely inevitable, much less threats that are more complex or over the horizon. There is, however, no inherent flaw in human nature or democratic government that keeps us from preparing for potential challenges and dangers by taking far-sighted actions with long-term benefits. As individuals, we do it all the time. The Congress did it in 1947. As a nation, today, as in 1947, the key is wise and focused bipartisan leadership—and political will.

I mentioned a moment ago that one of the most important lessons from our experience in Iraq, Afghanistan, and elsewhere has been the decisive role reconstruction, development, and governance plays in any meaningful, long-term success.

The Department of Defense has taken on many of these burdens that might have been assumed by civilian agencies in the past, although new resources have permitted the State Department to begin taking on a larger role in recent months. Still, forced by circumstances, our brave men and women in uniform have stepped up to the task, with field artillerymen and tankers building schools and mentoring city councils—usually in a language they don't speak. They have done an admirable job. And as I've said before, the Armed Forces will need to institutionalize and retain these non-traditional capabilities—something the ROTC cadets in this audience can anticipate.

But it is no replacement for the real thing—civilian involvement and expertise.

A few examples are useful here, as microcosms of what our overall government effort should look like—one historical and a few contemporary ones.

However uncomfortable it may be to raise Vietnam all these years later, the history of that conflict is instructive. After first pursuing a strategy based on conventional military firepower, the United States shifted course and began a comprehensive, integrated program of pacification, civic action, and economic development. The CORDS program, as it was known, involved more than a thousand civilian employees from USAID and other organizations, and brought the multiple agencies into a joint effort. It had the effect of, in the words of General Creighton Abrams, putting "all of us on one side and the enemy on the other." By the time U.S. troops were pulled out, the CORDS program had helped pacify most of the hamlets in South Vietnam.

The importance of deploying civilian expertise has been relearned—the hard way—through the effort to staff Provincial Reconstruction Teams, first in Afghanistan and more recently in Iraq. The PRTs were designed to bring in civilians experienced in agriculture, governance, and other aspects of development—to work with and alongside the military to improve the lives of the local population, a key tenet of any counterinsurgency effort. Where they are on the ground—even in small numbers—we have seen tangible and often dramatic changes. An Army brigade commander in Baghdad recently said that an embedded PRT was "pivotal" in getting Iraqis in his sector to better manage their affairs.

We also have increased our effectiveness by joining with organizations and people outside the government—untapped resources with tremendous potential.

For example, in Afghanistan the military has recently brought in professional anthropologists as advisors. The New York Times reported on the work of one of them, who said, "I'm frequently accused of militarizing anthropology. But we're really anthropologizing the military."

And it is having a very real impact. The same story told of a village that had just been cleared of the Taliban. The anthropologist pointed out to the military officers that there were more widows than usual, and that the sons would feel compelled to take care of them—possibly by joining the insurgency, where many of the fighters are paid. So American officers began a job training program for the widows.

Similarly, our land-grant universities have provided valuable expertise on agricultural and other issues. Texas A&M has had faculty on the ground in Afghanistan and Iraq since 2003. And Kansas State is lending its expertise to help revitalize universities in Kabul and Mazar-e-Sharif, and working to improve the agricultural sector and veterinary care across Afghanistan. These efforts do not go unnoticed by either Afghan citizens or our men and women in uniform.

I have been heartened by the works of individuals and groups like these. But I am concerned that we need even more civilians involved in the effort and that our efforts must be better integrated.

And I remain concerned that we have yet to create any permanent capability or institutions to rapidly create and deploy these kinds of skills in the future. The examples I mentioned have, by and large, been created ad hoc—on the fly in a climate of crisis. As a nation, we need to figure out how to institutionalize programs and relationships such as these. And we need to find more untapped resources—places where it's not necessarily how much you spend, but how you spend it.

The way to institutionalize these capabilities is probably not to recreate or repopulate institutions of the past such as AID or USIA. On the other hand, just adding more people to existing government departments such as Agriculture, Treasury, Commerce, Justice and so on is not a sufficient answer either—even if they were to be more deployable overseas. New institutions are needed for the 21st century, new organizations with a 21st century mindset.

For example, public relations was invented in the United States, yet we are miserable at communicating to the rest of the world what we are about as a society and a culture, about freedom and democracy, about our policies and our goals. It is just plain embarrassing that al-Qaeda is better at communicating its message on the internet than America. As one foreign diplomat asked a couple of years ago, "How has one man in a cave managed to out-communicate the world's greatest communication society?" Speed, agility, and cultural relevance are not terms that come readily to mind when discussing U.S. strategic communications.

Similarly, we need to develop a permanent, sizeable cadre of immediately deployable experts with disparate skills, a need which president bush called for in his 2007 state of the union address, and which the State Department is now working on with its initiative to build a civilian response corps. Both the President and Secretary of State have asked for full funding for this initiative. But we also need new thinking about how to integrate our government's capabilities in these areas, and then how to integrate government capabilities with those in the private sector, in universities, in other nongovernmental organizations, with the capabilities of our allies and friends—and with the nascent capabilities of those we are trying to help.

Which brings me to a fundamental point. Despite the improvements of recent years, despite the potential innovative ideas hold for the future, sometimes there is no substitute for resources—for money.

Funding for non-military foreign-affairs programs has increased since 2001, but it remains disproportionately small relative to what we spend on the military and to the importance of such capabilities. Consider that this year's budget for the Department of Defense—not counting operations in Iraq and Afghanistan—is nearly half a trillion dollars. The total foreign affairs budget request for the State Department is $36 billion—less than what the Pentagon spends on health care alone. Secretary Rice has asked for a budget increase for the State Department and an expansion of the Foreign Service. The need is real.

Despite new hires, there are only about 6,600 professional Foreign Service officers—less than the manning for one aircraft carrier strike group. And personnel challenges loom on the horizon. By one estimate, 30 percent of USAID's Foreign Service officers are eligible for retirement this year—valuable experience that cannot be contracted out.

Overall, our current military spending amounts to about 4 percent of GDP, below the historic norm and well below previous wartime periods. Nonetheless, we use this benchmark as a rough floor of how much we should spend on defense. We lack a similar benchmark for other departments and institutions.

What is clear to me is that there is a need for a dramatic increase in spending on the civilian instruments of national security—diplomacy, strategic communications, foreign assistance, civic action, and economic reconstruction and development. Secretary Rice addressed this need in a speech at Georgetown University nearly two years ago. We must focus our energies beyond the guns and steel of the military, beyond just our brave soldiers, sailors, Marines, and airmen. We must also focus our energies on the other elements of national power that will be so crucial in the coming years.

Now, I am well aware that having a sitting Secretary of Defense travel halfway across the country to make a pitch to increase the budget of other agencies might fit into the category of "man bites dog"—or for some back in the Pentagon, "blasphemy." It is certainly not an easy sell politically. And don't get me wrong, I'll be asking for yet more money for Defense next year.

Still, I hear all the time from the senior leadership of our Armed Forces about how important these civilian capabilities are. In fact, when Chairman of the Joint Chiefs of Staff Admiral Mike Mullen was Chief of Naval Operations, he once said he'd hand a part of his budget to the State Department "in a heartbeat," assuming it was spent in the right place.

After all, civilian participation is both necessary to making military operations successful and to relieving stress on the men and women of our armed services who have endured so much these last few years, and done so with such unflagging bravery and devotion. Indeed, having robust civilian capabilities available could make it less likely that military force will have to be used in the first place, as local problems might be dealt with before they become crises.

A last point. Repeatedly over the last century Americans averted their eyes in the belief that remote events elsewhere in the world need not engage this country. How could an assassination of an Austrian archduke in unknown Bosnia-Herzegovina effect us? Or the annexation of a little patch of ground called Sudetenland? Or a French defeat at a place called Dien Bien Phu? Or the return of an obscure cleric to Tehran? Or the radicalization of an Arab construction tycoon's son?

What seems to work best in world affairs, historian Donald Kagan wrote in his book On the Origins of War, "Is the possession by those states who wish to preserve the peace of the preponderant power and of the will to accept the burdens and responsibilities required to achieve that purpose."

In an address at Harvard in 1943, Winston Churchill said, "The price of greatness is responsibility … The people of the United States cannot escape world responsibility." And, in a speech at Princeton in 1947, Secretary of State and retired Army general George Marshall told the students: "The development of a sense of responsibility for world order and security, the development of a sense of overwhelming importance of this country's acts, and failures to act, in relation to world order and security—these, in my opinion, are great musts for your generation."

Our country has now for many decades taken upon itself great burdens and great responsibilities—all in an effort to defeat despotism in its many forms or to preserve the peace so that other nations, and other peoples, could pursue their dreams. For many decades, the tender shoots of freedom all around the world have been nourished with American blood. Today, across the globe, there are more people than ever seeking economic and political freedom—seeking hope even as oppressive regimes and mass murderers sow chaos in their midst—seeking always to shake free from the bonds of tyranny.

For all of those brave men and women struggling for a better life, there is—and must be—no stronger ally or advocate than the United States of America. Let us never forget that our nation remains a beacon of light for those in dark places. And that our responsibilities to the world—to freedom, to liberty, to the oppressed everywhere—are not a burden on the people or the soul of this nation. They are, rather, a blessing.

I will close with a message for students in the audience. The message is from Theodore Roosevelt, whose words ring as true today as when he delivered them in 1901. He said, "…as, keen-eyed, we gaze into the coming years, duties, new and old, rise thick and fast to confront us from within and from without…[The United States] should face these duties with a sober appreciation alike of their importance and of their difficulty. But there is also every reason for facing them with high-hearted resolution and eager and confident faith in our capacity to do them aright." He continued, "A great work lies ready to the hand of this generation; it should count itself happy indeed that to it is given the privilege of doing such a work."

To the young future leaders of America here today, I say, "Come do the great work that lies ready to the hand of your generation."

Thank you.

# Manama Dialogue

*Manama, Bahrain, Saturday, December 08, 2007*

THANK you very much. And my thanks to IISS for hosting not just this event, but other international dialogues as well—such as the Shangri La Dialogue in Singapore.

I would also like to express my gratitude to his majesty the king and to the people of Bahrain for their hospitality, and for their kindness to me and the U.S. delegation. For some six decades Bahrain—home to the U.S. Fifth Fleet—has been a valued friend and a contributor to security in the Gulf.

The partnership between Bahrain and the U.S. is one element of a broad American commitment to the security and stability of the Gulf, a commitment going back several decades and spanning multiple U.S. administrations.

The Gulf's security was a major focus of my previous service in government—first under President Carter, then President Reagan, and then the first President Bush.

- I remember the formulation of the "Carter Doctrine," based on the tenet that America would do what was necessary to defend our vital interests in the Gulf—a policy adopted by subsequent presidents that ultimately led to the creation of the United States Central Command;
- During the late 1980s, the U.S. stood with Bahrain and other members of the GCC to protect tanker ships and keep vital sea lanes open; and
- Then of course, the coalition that came together in 1990 to repel aggression against Kuwait.

This morning I would like to provide an overview of some issues that are of deep concern to the people of the United States and the people of the Middle East—issues that frame the security environment of the region now and almost certainly will for the foreseeable future. They include:

- The situation in Iraq today—both obstacles and opportunities;
- The implications of Iran's ongoing refusal to comply with its international obligations and the destabilizing effects of its actions and words; and third

- The ways the United States is strengthening our security ties in the region, with the aim of fostering partnerships amongst the nations of the Gulf.

It is useful to step back to when the Manama Dialogue last met, approximately one year ago, against a backdrop of deteriorating security and escalating sectarian violence in Iraq.

There was doubt in many quarters—at home and abroad—about whether the United States would be able to sustain our commitment in Iraq, and indeed in this part of the world broadly. The record of American activity over the past year should dispel that uncertainty. The United States remains committed to defending its vital interests and those of its allies in Iraq and in the wider Middle East.

I have just come from Iraq where I met both with American military commanders and Iraqi leaders. As many of you are no doubt aware, there has been in recent weeks and months a reduction in the level of violence and the return of a semblance of daily life in many cities and communities.

Attacks on Coalition and Iraqi troops have declined but, just as significantly, so have attacks against Iraqi citizens. Since the surge of U.S. forces began earlier this year, civilian deaths across Iraq are down about 60 percent, and they are down 75 percent in Baghdad. Recently, there was the lowest number of single-day attacks across the nation in three and a half years.

We attribute this to a number of factors, among them:

- A change of military tactics ordered by President Bush that emphasizes protecting the Iraqi people from insurgents, militias, and foreign terrorists;
- The increasing effectiveness of the Iraqi military, in tandem with our forces and in operations by themselves;
- The decision by some, but not all, militia groups to stand down from offensive operations; and
- The groundswell of ordinary citizens who have risen up to fight against Al Qaeda in Iraq and protect their families and their neighborhoods.

There is another side of the positive developments in Iraq, on the economic and civic front:

- The Saddam-era debt burden has been significantly reduced by international debt relief. I thank all nations that have contributed in this regard and urge remaining lenders to act promptly;
- Sound economic policies have led to low inflation rates, a stable currency, and a business environment that is more attractive to foreign investment. As a result, nominal growth is above 5 percent;
- While we are impatient that key legislation has not been passed, the presence of bottom-up reconciliation has had an impact. Provincial councils are more effective today than in the past, and concerned local citizens'

groups are giving more Iraqis a stake in the future of their nation. Local reconciliation across tribal, sectarian, and provincial boundaries is increasing the pressure at the national level to do likewise, as Iraq's leaders acknowledged to me;

- Our provincial reconstruction teams are facilitating these efforts by building capabilities for provincial councils, and working to improve the federal ministries. Both are efforts to improve community development, and both are gaining traction;
- And there are the intangibles—the beginnings of a return to normalcy and renewed hope that cannot be quantified by numbers.

The progress is real. But it is also fragile. The Iraqi government must use this breathing space bought with the blood of American, Coalition, and Iraqi troops to pass critical legislation:

- In order to solidify grassroots reconciliation;
- To improve the scope and effectiveness of government services;
- To signify that all elements of the government and society, whatever their beliefs and ethnicity, are, first and foremost, Iraqis; and
- Ultimately, to make the lives of all Iraqis better.

Iraq's government must demonstrate to its people, and to the world, that it can act with unity, purpose, and resolve.

Whether the positive trends of recent months continue will be largely determined by where we go from here. And by "we," I mean not only the Unites States and the Iraqi government, but also the governments of every nation represented at this dialogue.

It is no secret that U.S. troop levels in Iraq will begin to decline this month. That reality represents both risks and opportunities for the entire region.

As I said in Cairo earlier this year, whatever disagreements we may have over how we got to this point in Iraq, the consequences of a failed state—of chaos there—will adversely affect the security and prosperity of every nation in the Middle East and the Gulf region.

There may be some who, because of past resentments and disagreements, might be cheering for failure. I would respectfully suggest that these sentiments are dangerously shortsighted and self-destructive. The first and secondary effects of failure in Iraq—with all of its economic, religious, security, and geopolitical implications—will be felt in all the capitals and communities of the Middle East well before they are felt in the United States. The forces that would be unleashed—of sectarian strife, of an emboldened extremist movement with access to sanctuaries—do not recognize national boundaries and would surely target any government perceived to be a hindrance to their expansion of power.

Any nation that supports insurgents or militias in Iraq—either actively or passively—is in reality doing harm to itself, and to all of the people of the Middle East, be they Sunni, Shia, or any other sect. The primary victims of Al Qaeda and its affiliates in Iraq have not been Coalition troops or Iraqi security forces, but thousands of innocent civilians—men, women, and children whose only crime was to go to market or attend Friday prayers.

The terrorists have made their designs clear. Wherever they have seized territory in the past, they have made manifest their dark vision for the world. The future they seek is characterized not by piety or virtue, but by fear and intimidation. Not by martyrs holding true to their Islamic faith, but by criminals and suicide bombers whose only true allegiance is to murder and chaos.

But just as the nations of the Middle East have the most to lose from chaos in Iraq, they also have the most to gain from a secure, stable, and prosperous Iraq:

- An Iraq that is a net contributor to security in the Gulf;
- An Iraq that will be a strong and vibrant trading partner;
- An Iraq with the potential to set an example of responsive, effective governance; and
- An Iraq that helps bridge the sectarian divides in this part of the world.

I urge you to exercise your influence with the Iraqis and encourage them to meet their own goals and expectations, to live up to their own promises. I also urge you to help them in every way you can—by dampening homegrown insurgencies, by alleviating sectarian strife, by providing economic and diplomatic support. Iraq is a multi-ethnic Arab state, like many in the region. For other Arabs to withhold support and friendship because of the composition of Iraq's government, and as that government determines its future path at home and with its neighbors, is to increase the risk of the very outcome many in the region fear.

Iraq is not an island, and its future is closely tied to the behavior of its neighbors—for better or for worse. Which brings me to Iran.

For 29 years I have been watching the Iranian government intently, at the Central Intelligence Agency, at the National Security Council, and now at the Department of Defense. My first direct encounter with the leaders of the revolutionary government of Iran was in Algiers in October 1979, just eight months after that government came to power. For 29 years, I have followed this government's words, its promises—mostly hollow—and, more tellingly, its deeds, both overt and covert.

This week, however, marks a watershed. Astonishingly, the revolutionary government of Iran has this week, for the first time, embraced as valid an assessment of the United States intelligence community—on Iran's nuclear

weapons program. And since that government now acknowledges the quality of American intelligence assessments, I assume that it will also embrace as valid American intelligence assessments of:
- Its funding and training of militia groups in Iraq;
- Its deployment of lethal weapons and technology to both Iraq and Afghanistan;
- Its ongoing support of terrorist organizations—like Hezbollah and Hamas—that have murdered thousands of innocent citizens; and
- Its continued research and development of medium-range ballistic missiles that are not particularly cost-effective unless equipped with warheads carrying weapons of mass destruction.

In reality, you cannot pick and choose only the conclusions you like of this recent National Intelligence Estimate. The report expresses with greater confidence than ever that Iran did have a nuclear weapons program—developed secretly, kept hidden for years, and in violation of its international obligations. It reports that they do continue their nuclear enrichment program, an essential long lead time component of any nuclear weapons program. It states that they do have the mechanisms still in place to restart their program. And, the estimate is explicit that Iran is keeping its options open and could restart its nuclear weapons program at any time—I would add, if it has not done so already. Although the Estimate does not say so, there are no impediments to Iran restarting its nuclear weapons program—none, that is, but the international community.

Everywhere you turn, it is the policy of Iran to foment instability and chaos, no matter the strategic value or the cost in the blood of innocents—Christians, Jews, and Muslims alike. There can be little doubt that their destabilizing foreign policies are a threat to the interests of the United States, to the interests of every country in the Middle East, and to the interests of all countries within the range of the ballistic missiles Iran is developing.

Considering all this, the international community should demand that the Iranian government come clean about the extent of its past illegal nuclear weapons development. The international community should insist that Iran suspend enrichment. The international community should require that the Iranian government openly affirm that it does not intend to develop nuclear weapons in the future and, further, that it agree to inspection arrangements that will give us all confidence that it is adhering to that commitment.

Let us be realistic.

Although our nations have differing perspectives and histories, we nonetheless share a deep concern about Iran's current course. While we must keep all our options open, the United States and the international community must continue—and intensify—our economic, financial, and diplomatic pressures on Iran to suspend enrichment and to agree to verifiable arrangements that can prevent that country from resuming its nuclear weapons program at a moment's notice—at the whim of its most militant leaders. That should be a matter of grave concern to every government in the world. Let us continue to work together to take the necessary peaceful but effective measures necessary to bring a long-term change of policies in Tehran.

The challenges that I have discussed—and others—require that civilized nations—in the Gulf and worldwide—work together in ways that previously may have seemed unnecessary, or even unwelcome, but that are an absolute necessity today.

In the past, bilateral arrangements with the United States have helped maintain a balance of power in the Gulf region. While such partnerships are important, the United States seeks to encourage more multilateral ties and cooperation with and among our friends in the region.

In October, we began the third round of talks in the Gulf Security Dialogue, a strategic framework designed to enhance and strengthen regional security—which includes strengthening Iraq. The Gulf Security Dialogue helps counter conventional as well as unconventional, asymmetric, and terrorist threats by focusing on six key pillars:

- Defense cooperation;
- Developing a shared assessment and agenda on Iraq;
- Regional stability, especially with respect to Iran;
- Energy infrastructure security;
- Counter-proliferation; and
- Counter-terrorism.

I would like to expand briefly on the first pillar—defense cooperation—which may be of particular interest to this audience. Defense cooperation discussions in the Dialogue specifically addressed issues such as shared early warning, cooperative air and missile defense, and maritime security awareness.

Initiatives like the Bilateral Air Defense Initiative could become a stepping stone to a multilateral effort to develop regional air and missile defense systems that would provide more comprehensive coverage, a regional protective—defensive—umbrella. We should bear in mind the deterrent effect such a system would have. If the chances of a successful attack are greatly reduced, then so too is the value of pursuing offensive weapons systems and delivery systems.

Maritime security awareness might include developing a maritime surface picture and standard operating procedures against seaborne threats, such as terrorism, piracy, narcotics trafficking, and smuggling.

All told, the Dialogue will bolster the defensive capabilities of nations in the region, while not diminishing our bilateral relationships or U.S. commitments in the region. This framework will also enhance interoperability between our armed forces and foster cooperation among participating nations through exercises and training. As the U.S. Secretary of State said last July, "Through our Gulf Security Dialogue, we are helping to strengthen the defensive capabilities of our partners ... [to] support their ability to secure peace and stability in the Gulf region."

The Gulf Security Dialogue is not intended to replace ongoing bilateral and multilateral efforts, but rather to enhance such efforts by building the capacity of gulf nations to deal with common threats in partnerships with the United States.

Two weeks ago, representatives from 49 nations gathered in Annapolis, where Israeli and Palestinian leaders agreed to begin negotiations toward achieving a lasting peace. Some people have criticized the Annapolis summit because of what it represented—a process, or the beginning of a process, as opposed to immediate results. I believe this view is shortsighted.

During the 1970s, many people discounted the value of holding strategic talks with the former Soviet Union—America's most dangerous adversary—because these meetings often did not lead directly to new arms control breakthroughs. It turned out that maintaining the dialogue helped each side better understand the other's intentions, and laid the groundwork for gains that ultimately brought the Cold War to a close. With persistence, courage, and good faith on both sides, I believe it will be possible to see what President Bush called "the expansion of freedom and peace in the Holy Land."

I believe that through these and other initiatives, with the right leadership from the countries represented in this room, there will arise new possibilities for the citizens of the Middle East. After all, we are living at a time where, as never before, people here and around the globe are demanding and making progress toward peace, political openness, and an economic system that works for themselves and their families.

For all of the difficulties we now confront—and they are daunting, to be sure—I believe that, over time, there can be a very different future—a different narrative, if you will—for this part of the world. A future:

- Where trade, commerce, and economic opportunity lead to a growing middle class and a higher quality of life for workers and their families;
- Where Palestine and Israel are living in peace side by side as viable, secure, and independent states;
- Where men and women have an increasingly greater say and a greater stake in how they govern their own lives, their own communities, and their own countries; and
- Where citizens from Tehran to Baghdad to Beirut can look forward to a life secure from the assassin, the suicide bomber, and the proverbial knock on the door in the middle of the night.

Reaching these goals cannot be achieved by any one nation alone—no matter how wealthy or powerful. And they certainly will not be achieved by military means alone.

That said, there may be some in the region who believe that the staying power and strength of the United States have been diminished or undermined by the wars in Iraq and Afghanistan. They may believe our resolve has been corroded by the challenges we face at home and abroad.

This would be a grave misperception. Over the past century, many nations and empires and movements have looked to our shores in search of signs of vulnerability—signs that Americans are weak or undisciplined; that we are stretched thin and unable to fulfill our commitments; that we do not have the patience or the will to face a long-term challenge; that open and vigorous debate in our democracy reflects underlying divisions and irresolution with respect to defending our vital interests.

Imperial Germany, Imperial Japan, Nazi Germany, Fascist Italy, the Soviet Union—all made this fundamental miscalculation. All paid the price. All are on the ash heap of history.

As I have said before, restraint should never be confused with weakness.

Surveying the overall trajectory of the Middle East, I believe the greatest challenges we face will ultimately be decided by the choices we make—choices faced by all of the governments represented here:

- Between the divisions and rancor of the past, and the strength and stability that comes from reconciliation and cooperation;
- Between proxy wars that victimize innocents, and support for the brave men and women who are struggling to rebuild and secure their homelands;
- Between safe detachment from the plight of one's neighbors, and working to improve stability and prosperity for everyone in the region.

As we consider the security challenges of today, we must be ever cognizant that the choices we make will for many years weigh heavily on the fate of our peoples. To meet great expectations, we must all be willing to take risks—for peace, for security, for the future of our children.

Thank you.

# House Armed Services Committee
# Hearing on Afghanistan

*Capitol Hill, Washington, DC, Tuesday, December 11, 2007*

SUBMITTED Statement for the Record

Mr. Chairman, Representative McHugh, members of the committee, thank you for inviting us to testify before you today. We have a—I have a longer statement that we've submitted for the record. As you've noted, I've just returned from Afghanistan, where I met with Afghan officials, U.S. commanders, our civilian colleagues, and our European allies. And this is an opportune time to discuss our endeavors in that country.

I would tell you that when I took this job, it seemed to me that the two highest priorities that I had were our wars in Afghanistan and in Iraq. If I'm not mistaken, I just finished my sixth trip to Iraq and, I think, maybe my fourth trip to Afghanistan.

Notwithstanding the news we sometimes hear out of Afghanistan, the efforts of the United States, our allies and the Afghan government and people have been producing some solid results. If I had to sum up the current situation in Afghanistan, I would say there is reason for optimism, but tempered by caution.

Projects that will have a real impact on the lives of citizens are under way, with the construction of utilities, roads and schools. The Congress has appropriated about $10 billion in security and reconstruction assistance to Afghanistan for fiscal year 2007—almost three times the previous year's appropriation. I thank you, the members of Congress, for your strong support of this effort. Admiral Mullen will speak in more detail about some of the activities made possible by the funding increase with regard to Provincial Reconstruction Teams, Afghan security force, as well as our own endeavors.

We've just passed the first anniversary of NATO's taking over all responsibility for helping Afghans secure their democracy.

The first half of 2007, NATO and coalition forces took the initiative away from the Taliban. Contributions from our civilian colleagues helped secure these military gains. Afghan forces played a key role, demonstrating their improved capability in the last year, and indeed Afghan security forces have led the fight to retake Musa Qal'eh in recent days.

As you know, in 2007, the number of terrorist attacks in Afghanistan increased. The insurgents have resorted more and more to suicide bombs and improvised explosive devices similar to those found in Iraq. As I learned during my visit, some of the uptick can be attributed to increased Afghan and ISAF operations. The Taliban and their former guests, al Qaeda, do not have the ability to reimpose their rule, but only in a truly secure environment can reconstruction projects take root and rule of law be consolidated. That environment has not yet been fully achieved, but we are working toward it.

As you know, the drug trade continues to threaten the foundations of Afghan society and this young government. To attack this corrosive problem, a counter-narcotics strategy is being implemented that combines five pillars: alternative development, interdiction, eradication, public information and reform of the justice sector. I hope that the coming year will show results.

There also needs to be more effective coordination of assistance to the government of Afghanistan. A strong civilian representative is needed to coordinate all nations and key international organizations on the ground. We and others have worked with the Karzai government to identify a suitable candidate. I'm hopeful this exhaustive search will be completed soon.

The final point I'll turn to—and it is an extremely important one, and both you, Mr. Chairman, and Mr. McHugh referred to this—is the willingness of our NATO allies to meet their commitments.

Since ISAF assumed responsibility for all of Afghanistan in October 2006, the number of non-U.S. troops has increased by about 3,500. NATO still has shortcomings—shortfalls in meeting minimum requirements in troops, equipment and other resources. I leave for Scotland tomorrow for a meeting of defense ministers of the countries involved in Regional Command South, and this will certainly be on the agenda.

The Afghanistan mission has exposed real limitations in the way the alliance is or organized, operated and equipped. I believe the problem arises in a large part due to the way various allies view the very nature of the alliance in the 21st century, where in a post-Cold War environment, we have to be ready to operate in distant locations against insurgencies and terrorist networks.

I would also like to stress the role Congress can play in this endeavor. If other governments are pressured by this body and by the Senate as well as by those of us in the executive branch, it may help push them to do the difficult work of persuading their own citizens of the need to step up to this challenge.

Let me close by telling you about a region I visited last week, a region that demonstrates why I am cautiously hopeful about the mission in Afghanistan. For years, and even decades, the Khowst region has been a hotbed of lawlessness and insurgent activity.

Things are very different today. Under the strong leadership of an honest and capable governor, and with Afghans in the lead, there have been remarkable gains as security forces, local organizations and the U.S.-led Provincial Reconstruction Team, with representatives from the State Department, USAID and the Department of Agriculture, have worked in tandem to promote civic and economic and development. Where last year there was one suicide bombing per week, now there is on average one per month.

As the governor said to me, through our combined efforts more has been accomplished in the past eight months than in the prior five years.

Khowst is a model of the integration of hard and soft power in a counterinsurgency campaign. And it is an example of what can be done in other regions.

You have asked us to talk about the way forward. I would tell you that I proposed at the last NATO defense ministerial that NATO put together a strategic concept paper looking forward three to five years; where do we want to be in Afghanistan and what will be the measures of progress? We will be talking about that in Scotland over the next couple of days. The rest of the alliance defense ministers have embraced this idea, and my hope is that we can present such a strategic concept paper to the heads of state at their meeting in Bucharest next spring.

A moderate, stable Afghanistan is crucial to the strategic security of the United States and its allies. The elected leaders of the countries that make up our alliance have said as much. Afghans have the will to keep their nation in the democratic fold, and we need to match their determination with the necessary resolve and resources to get the job done.

# 23rd Annual Martin Luther King, Jr. Observance

*Washington D.C., Thursday, January 17, 2008*

THANK you, Mike. Thank you all for being here today.

This is a day to remember a great American. It is also a day to reflect on what we can do to further the struggle for human freedom and dignity that Dr. King helped lead and for which he gave his life.

Dr King pushed the country to adhere to the just and true idea on which it was founded: that all human beings are equal in their God-given right to life, liberty, and the pursuit of happiness. In fighting for an end to racial discrimination, he used tactics that showed how well he understood the nation he sought to change for the better.

He said: "The only weapon that we have in our hands ... is the weapon of protest."

Nonviolent protest was a tactic that could not be employed just anywhere. At the Montgomery bus boycott in 1955, Dr. King said: "If we were incarcerated behind the iron curtains of a communistic nation, we couldn't do this. If we were trapped in the dungeon of a totalitarian regime, we couldn't do this. But the great glory of American democracy is the right to protest for right."

It is easy to forget, at this distance in time, that change was needed everywhere, not just in the south.

In 1958, when I was 14, nonviolent protest came to my home town of Wichita, Kansas. There was a drug-store chain called Dockum, and at one of its Wichita stores that summer, a group of some 20 black teenagers and young adults sat down for service at the lunch counter. They were refused. They kept coming back for three weeks, endured many indignities, until the management changed its policy. This bold action by the local youth chapter of the NAACP was the first successful, student-led sit-in of its kind. It helped end segregation at drug stores throughout Kansas, pre-dating by two years the more famous sit-ins at a North Carolina Woolworth's.

I now have the honor of leading an institution that began breaking down the barriers of race at the dawn of the modern civil rights revolution. African Americans have represented the United States with honor and distinction.

In recent years, they have participated in the defense of the nation well beyond their percentage of the population.

Our featured speaker today, Lieutenant General Michael Rochelle, has charge of the United States Army's personnel system at a crucial time. The force needs to expand, and General Rochelle, as Army's G-1, is overseeing the first significant increase in a generation. It is a tall order—to grow the force in a way that relieves the stress from current military operations, enables the United States to meet its commitments at home and abroad, and achieves these goals without sacrificing the quality we have come to expect in our all-volunteer force.

I have every confidence in him. The Army and the nation are depending on General Rochelle, and my hope and expectation is that, in the years ahead, more African Americans will staff the Army and other branches at the highest levels following the examples set by Generals Colin Powell, Kip Ward, and many others.

General Rochelle has often spoken of mentoring young people to encourage minorities and women to excel, and to grasp the opportunities for advancement that the armed forces provide. As Dr. King said, "human progress never rolls in on wheels of inevitability." It takes "people of goodwill put[ting] their bodies and their souls in motion."

Everyone at the department must be sensitive to the need to build these mentoring relationships, and must act to make sure that this is taking place in every service, at every level.

Only if we make this concerted effort will our military continue to be the greatest equalizing institution in the United States of America. Let us keep pushing for progress. We must keep pushing for progress. In the words of Martin Luther King, "Let us march to the realization of the American dream."

Thank you all very much, and thank you for being here this morning.

## Space and Naval Warfare Center MRAP Facility

*Charleston, SC, Friday, January 18, 2008*

THANK you. This is indeed an awesome operation, and I'll only interrupt your work for about five minutes. To the men and women of SPAWAR, I wanted to take a few moments to thank you for your hard work and dedication to this important effort. MRAP is a proven lifesaver on the battlefield. You have my appreciation and my respect—but more importantly, the thanks of countless moms and dads, husbands and wives, and sons and daughters of U.S. troops deployed abroad.

As you know, IEDs are the tactic of choice for our enemies. They are cheap, and deadly, and difficult to detect on the dusty streets of Baghdad, Samarra, Mosul, and elsewhere. They have been the biggest killer of our troops in Iraq.

There is no failsafe measure that can prevent all loss of life and limb on this or any other battlefield. That is the brutal reality of war. But vehicles like MRAP, combined with the right tactics, techniques, and procedures, provide the best protection available against these attacks.

Last year, I made MRAPs the Defense Department's top acquisition priority. By the end of 2007, the target of building, integrating, and delivering 1,500 fully capable MRAPs was met—because you worked six days a week and around the clock. I'm told that nearly a third of those who work here at SPAWAR are veterans, who know from experience how important it is to get the best equipment to the battlefield as soon as possible.

The partner manufacturers and suppliers, and you here at SPAWAR, have delivered under pressure with lives on the line. In fact, the last time American industry moved from concept to full-rate military production in less than a year was World War II.

This has been a team effort with many moving parts—in the military and industry, elsewhere in the private sector. Suppliers of steel, tires, and other materials have stepped up, as have the manufacturers—firms in the United States and in ten foreign countries. Through the efforts of Transportation Command, MRAPs are shipped or flown from here in Charleston to places halfway around the world.

I don't think it will surprise you to hear me say you must keep pressing on. IEDs will be with us for some time to come—in Iraq, Afghanistan, the battlefields of the future. The need for these vehicles will not soon go away.

The war effort of the 1940s mobilized the entire American economy. That is not the case today, and while that may make yours seem like a lonely task, it only underscores how important that task is. Back then, President Roosevelt said: "We must raise our sights all along the production line. Let no [one] say it cannot be done." Those in the MRAP program have shown that it can be done. So keep raising your sights. Keep these vehicles rolling off the line. Your efforts are saving lives. Of the MRAPs newly in the hands of the Army in Iraq, 12 have been attacked. Every soldier walked away.

To put it in the words of one Sergeant Major, and I quote, "MRAP is just lovely!" Actually, he didn't say "lovely." He was actually considerably more colorful than that in his comments. But I will quote the rest of his statement. He said, "Troops love them, commanders sleep better knowing the troops have them." There can be no better description of the difference you are making here. You are saving lives.

So just as I am thanking you now, America and our men and women in uniform have great reason to thank you as well—now, and in the years to come. Thank you.

## America Supports You Summit

*Pentagon, Friday, January 25, 2008*

ALLISON, thank you for that introduction. On behalf of the entire Department of Defense, let me thank you all for all you do and have done to support our troops.

Brad Avery, I understand you recently returned from Iraq as part of a USO tour. Thank you for going. And I'm sure everyone here is looking forward to your performance.

I have to admit, this is the first time I've ever been a warm-up act for a member of a Grammy Award winning band. I can add a little embarrassment with Bono visiting the other day. I thought U2 was a plane. I appreciate the warm welcome. Too often when I'm asked to attend events here in Washington, the first thing my hosts do is often ask me to please raise my right hand. And then ask if I'm actually going to tell them the truth.

Of course, you expect that kind of thing in Washington. A city where those who travel the high road of humility encounter little heavy traffic. The only place in the world you can see a prominent person walking down lover's lane holding his own hand. Where people say, "I'll double-cross that bridge when I get to it."

As Allison said, I've been at the Pentagon for just over a year—which brings to mind President Truman's comment about Washington. He said, "For the first six months, you wonder how the hell you got here. For the next six months, you wonder how the hell the rest of them ever got here."

Happily, I think your experience today has been quite different. And I hope you've had a chance to see the Pentagon and learn more about the Department. I hope we were able to provide you with some new tools and ideas that will help you with your mission.

During my time as Secretary of Defense, I have had the opportunity to travel all over the world and meet with our men and women in uniform, at every level, from privates to four-stars. I've also met with the families of these extraordinary service members. In all of these encounters, I feel honored to serve alongside them; humbled by their extraordinary sense of duty and dedication and their willingness to risk life and limb in defense of our nation; and blessed to live in a country with so many brave men and women.

As Americans, we owe them so much. And as only Americans can do, I believe that our citizens—you—have risen to the occasion—a far cry from the last time we were engaged in a controversial war in the 1960s and 1970s. You see it in airports all over the country, where soldiers are met with standing ovations by passengers in the terminal. I've been there, I've seen it myself. There are free meals and rounds of drinks. And, above all, simple thank yous. The appreciation is real, it is heartfelt, and it bridges any political divide.

And while we're all united in our admiration of those who have volunteered to serve our nation during these challenging times, it takes a special kind of person to devote part of their life to actively making the lives of our troops better—both during their deployments and when they get back.

Looking over the groups represented here at this America Supports You summit, I was struck by a number of things—the diversity of the geographic areas you hail from; the range of activities you do to support our troops; and the collective magnitude of what has been accomplished by citizens like you. Citizens who have felt a call to support your fellow Americans in a time of need, with gestures large and small. Whether that is:

- Sending care packages to troops or providing school supplies and toys for Iraqi children;
- Creating education scholarships and finding job opportunities;
- Helping veterans navigate the maze of government bureaucracy;
- Supporting the families of the fallen and those recovering from injuries;
- And countless other activities to improve the lives of men and women who have served our nation with honor and distinction.

For those whose lives you have touched, every gesture, no matter how small, has a tangible impact. Your work plays a vital role in uplifting spirits in the face of dangers and stresses on the battlefield and at home.

It is noble work—done not to garner publicity or get invited to the Pentagon. You do it because you feel, like I do, a deep pride in a new generation of Americans who, when faced with extraordinary challenges, have answered a call to duty, honor, and country.

These days we hear a lot about how our society for the most part is not involved in the war effort—that most citizens are not directly affected by the ongoing conflicts. There is an element of truth to those claims. But in making them, there is also a tendency to overlook all the good work that is being done on behalf of our troops—the work being done by organizations such as yours and by compassionate and selfless citizens across the nation.

And so, from the bottom of my heart, and on behalf of all our men and women in uniform and their families, thank you for everything you have done and will continue to do in the future.

Thank you.

## Center for Strategic and International Studies

*Washington, DC, Saturday, January 26, 2008*

THANK you for the kind comments, John [Hamre], and, Sam [Nunn], thank you for your kind remarks as well. I should note that John has had the courage if not necessarily the good judgment to accept my invitation to chair the Defense Policy Board. And I thank him for it.

Well, it's a real pleasure to be back in Washington—not—— the place where Harry Truman said you spend the first six months wondering how the hell you got here and the next six months wondering how the hell the rest of them got here. To quote Senator Alan Simpson, "Washington is the only place where those who travel the high road of humility encounter little heavy traffic." The only place in the world you can see a prominent person walking down Lover's Lane holding his own hand.

Well, looking out to the audience today, I see no shortage of friends, colleagues, and distinguished figures from the worlds of politics, diplomacy, business, the military, and academia. I'm sure that conspiracy theorists would have a field day with this gathering, right up there with the trilateral commission and fluoridated water. I have to say you begin to think in those terms when you've had a guy wearing a football helmet covered with tin foil outside the White House for a year holding a sign accusing you of controlling his brain waves.

There are many friends here today, but I would like to just indulge myself by singling out three people: first, Dave Abshire, CSIS co-founder, still going strong and still bettering our understanding of statecraft and the importance of integrity and leadership; Anne Armstrong, former ambassador to the United Kingdom, stalwart supporter of CSIS and perhaps most important to me former member of the board of regents of Texas A&M. Had it not been for Anne's support, I would not have been selected as president of A&M; And, finally, George Schultz. As I wrote in my book more than 10 years ago, I believe history will record George as one of America's greatest Secretaries of State.

Two months ago, giving the Landon lecture at Kansas State University, I made the case for increasing the capacity of America's civilian tools of statecraft and for better integrating them with the hard power of our military, or as John Hamre once put it, to combine the tools of "intimidation" with the tools of "inspiration," also called smart power. Of course, what got the media's attention was the man-bites-dog aspect of the speech, the Secretary of Defense calling for a significant increase in the budget of the Department of State.

I dare say, Secretary Schultz, I suspect that did not happen during your tenure. And so, John asked me to talk about this subject here today. In recent years, we have seen that the close of the Cold War, an event that raised hopes for an era of prosperity, tranquility, and, quote, unquote, "the end of history" also had the effect of unfreezing ancient hatreds and unleashing new pathologies. The revived monsters of the past have returned far stronger and more dangerous than before because of modern technology, both for communication and for destruction and to a world that is far more closely connected and interdependent.

For years to come, we will deal with a new, far more malignant form of global terrorism rooted in extremist and violent jihadism, new manifestations of ethnic, tribal, and sectarian conflict, the proliferation of weapons of mass destruction, failed and failing states, states enriched with oil profits and discontented with their place in the international system, authoritarian regimes facing increasingly restive populations that seek political freedom as well as a better standard of living, and, finally, we see both emergent and resurgent great powers whose future paths remain unclear.

These challenges have two things in common. First, they are, by their nature, long term, requiring patience over years and across multiple presidencies. Second, they cannot be overcome by military means alone and they extend well beyond the traditional domain of any single government agency or department. They require our government to operate with unity, agility, and creativity, and will require devoting considerably more resources to nonmilitary instruments of national power.

In the Afghanistan and Iraq campaigns, one of the most important lessons that has been learned, and to a large extent, relearned is that military success is not sufficient. Our efforts must also address economic development, institution building, the rule of law, promoting internal reconciliation, good or at least decent governance, public services, training and equipping indigenous security forces, effective, strategic communications, and more. These so-called soft capabilities along with military power are indispensable to any lasting success, indeed, to victory itself as Clausewitz understood it, which is achieving a political objective.

Despite the heroic effort of individual soldiers and diplomats and many successful operations—the surge in Iraq being the most recent and compelling example—the whole of our government's activities has often added up to less than the sum of the parts.

The military and civilian elements of our national security apparatus have responded unevenly and have grown increasingly out of balance. For example, in Iraq and Afghanistan, soldiers, sailors, Marines, and airmen have filled the void created by the absence of civilians available to deploy and operate in different and dangerous environments. I dealt with this shortly after becoming Defense Secretary when the Iraq surge was announced.

A key part of the plan was more provincial reconstruction teams. The Department of Defense soon thereafter thereafter received a memo from State asking for military personnel to fill the civilians slots on the PRTs. As you might imagine, this provoked a somewhat negative response in Defense.

But the problem is not will; it is capacity. In many ways, we are still coping with the consequences of the 1990s, when with the complicity of both ends of Pennsylvania Avenue, key instruments of American power abroad were reduced or allowed to wither on the bureaucratic vine. The State Department froze a hiring of new Foreign Service officers for a period of time. The U.S. Agency for International Develop dropped from a high of 15,000 permanent staff during the Vietnam War to about 3,000.

And then there was the U.S. Information Agency. At one point, its directors included the likes of Edward R. Murrow. It was split into pieces and folded into a corner of the State Department. Since September 11th, and through the efforts of first, Colin Powell, and, now, Condi Rice, the State Department has made a comeback. Foreign Service officers are being hired again and foreign affairs spending has about doubled since President Bush took office.

But shortfalls persist. A couple of weeks ago, I spoke with several dozen U.S. ambassadors who were visiting Washington for a Chiefs of Mission conference. The speaker who preceded me on the program was the Director General of the Foreign Service, the State Department's chief personnel officer. I'm told that his briefing was sobering, bordering on grim, of unfilled billets across the word due to shortages of mid-level and senior-level officers, caused by earlier hiring freezes and the staffing requirements of Iraq. Additionally, about 30 percent of AID's foreign service officers are eligible for retirement, valuable experience that cannot be contracted out.

This is why I believe we need to think about America's investment in foreign affairs on a fundamentally different scale. It is useful to remember that the amount of national treasure it would take to fund a major boost in civilian capabilities is relatively small. In a week and a half, I will go to Capitol Hill to present the fiscal year 2009 Defense budget. It's no big secret that the total will be somewhere around a half a trillion dollars. The total foreign-affairs request last year was $36 billion, about what the Pentagon spends on health care. Another comparison—the Army is planning to add about 7,000 more soldiers in 2008 to the active Army. It's part of a multi-year expansion. In pure numbers, that is equivalent to adding the entire U.S. Foreign Service to the Army in one year.

Beyond filling current voids in staffing and operations, a permanent sizable increase in the ranks of foreign service, if done properly, would have significant institutional benefits in terms of State's capacity and its influence vis-à-vis other agencies.

To give you a military example, a certain percentage of officers, even in time of war and when the force is stretched, are always enrolled in some kind of advanced training and education and leadership, strategy, or planning at the staff and war colleges and at graduate school. No such float of personnel exists for the Foreign Service. The same is true of planning. Between the joint staff, the services, and various commands, the military has thousands of officers dedicated to planning in some form. That kind of capacity does not exist on the civilian side of the government.

Despite the relatively modest amounts of money involved, getting the additional resources and authorities for soft power is not an easy sell politically. It simply does not have the built-in, domestic constituency of defense programs. As an example, the F-22 aircraft is produced by companies in 44 states; that's 88 senators. However, within the senior ranks of the military, a real constituency does exist for strengthening the non-military tools of national power. Admiral Mike Mullen, the Chairman of the Joint Chiefs of Staff once said as Chief of Naval operations that he would hand over a portion of his budget to the State Department in a heartbeat, assuming it was spent in the right place.

After all, civilian participation is necessary to the success of most military operations. As we have seen in PRTs and elsewhere, the inclusion of even a few properly employed civilian experts becomes what the military calls force multipliers

But we have to be realistic about how much even well-funded civilian agencies can do to reduce the demands on our armed forces to conduct what in recent years has been called non-traditional missions. Ever since General Winfield Scott led his army into Mexico in the 1840s, virtually every major deployment of American force has led to a longer military presence to maintain stability. General Eisenhower, when tasked with administering North Africa in 1942, wrote, "The sooner I can get rid of all of these questions that are outside the military in scope, the happier I will be! Sometimes, I think I live 10 years each week, of which at least nine are absorbed in political and economic matters."

During World War II, the Army even established a school of military government whose students played a key role in post-war Germany and Japan. And after much of the military establishment said "never again" following Vietnam, U.S. Armed Services found themselves again policing and rebuilding places like Somalia, Haiti, the Balkans, and, now, Afghanistan and Iraq. The requirement for the U.S. military to maintain security, provide aid and comfort, begin reconstruction, and stand up local government and public services will not go away. At least in the early phases of any conflict, military commanders will no more be able to rid themselves of these tasks than Eisenhower was.

As a former U.N. Secretary General once said about peacekeeping, "It is not a job for soldiers, but only soldiers can do it." I told an Army gathering last year that it is hard to conceive of any country challenging the United States directly in conventional military terms for some time to come. We can expect these so-called asymmetric operations, messy, protracted struggles without clear battle lines or exit strategies to be a mainstay of the 21st century battlefield.

So the military must retain the lessons and institutionalize the capabilities it has learned and relearned in these key areas. The military and our government as a whole is grappling with the reality that the fundamental nature of conflict as we've long perceived it has changed. As we have seen from the recent campaigns, the once stark black-and-white divisions between war and peace have faded. And so, America's national security apparatus, military and civilian, needs to be more adept in operating along a continuum involving military, political, and economic skills in a gray area that is likely to be persistent, containing opportunities as well as dangers. These scenarios will call for more shaping and influencing and less compulsion of friends, adversaries, and, most importantly, those in between.

Over the past 15 years, we have tried to overcome post-Cold War challenges and pursue 21st century objectives with processes and organizations designed in the wake of the Second World War. The National Security Act that created most of the current interagency structure was passed in 1947. The last major legislation structuring how American dispenses foreign aid was passed during the Kennedy administration. The U.S. government has tried, incrementally, to modernize our posture and processes in order to improve interagency planning and cooperation mostly through a series of new directives, offices, coordinators, tsars, and various initiatives.

And there are some signs of progress.

- Two years ago, Secretary Rice initiated was has been called transformational diplomacy. She questioned why the United States had as many diplomats stationed in Germany, a nation of 82 million people, as in India with a billion people. People and resources are being moved from where they made sense during the Cold War to where they make sense now.

- At last year's State of the Union, President Bush called for a new civilian reserve corps in the State Department, a permanent, sizeable corps of deployable experts comparable to the National Guard in the military arena.

243

- New joint authorities have been granted by Congress to State and Defense that allow us together to train and equip partner security forces with more speed and flexibility than in the past.
- The number of civilians deployed in PRTs has increased over the past year. And the State Department recently announced that it is doubling the number of Foreign Service officers assigned to military headquarters in the United States and abroad. In fact, one of the two deputy commanders of the new Africa Command will be a State Department officer.
- A new executive order on national security professional development encourages Foreign Service officers and civil servants from State as well as the military and other departments to serve tours in other agencies in a way that enhances their career and promotion prospects.

We are also looking toward more untapped resources outside of the government, places where it's not necessarily how much you spend, but how and where you spend it. After World War II, the defense establishment realized it needed to be better connected to the academic and scientific communities, not only for new technology and weaponry but for their insights into history, strategy, and economics. And this led to the creation of institutions like RAND.

We are once again trying to mine these resources for cultural expertise. Over the past year, for example, the military has been advised by anthropologists in Afghanistan called human terrain teams. In one case, the anthropologist pointed out that in one Afghan village, there were many widows, and that their sons might feel compelled to take care of them by joining the Taliban where many of the fighters are paid. And so, the American officers started a job training program for the widows. Similarly, American land grant universities like Texas A&M have deployed teams to Iraq and Afghanistan that provide valuable expertise in agriculture and other areas.

As in any new venture, there has been resistance. The human terrain teams have met with some pushback in academia where the military is sometimes viewed with a certain measure of suspicion. But it is important that we take advantage of the expertise available outside the ranks of the government.

As important and promising as many of these initiatives are, they have often been created ad hoc and on the fly in a climate of crisis. We need to figure out how to institutionalize and integrate programs such as these. The ultimate answer is probably not going to be recreating the old USIA or AID or simply adding more deployable people to State and other agencies. New approaches and new institutions are required for the 21st century.

Looking forward, bureaucratic barriers that hamper effective action should be rethought and reformed. The disparate strands of our national security apparatus, civilian and military, should be prepared ahead of time to deploy and operate together.

I should note that some of these challenges are not new. Our government has always been plagued by turf wars and stovepipes and conflicts over personality and ideology. During the Cold War, there were military, intelligence, and diplomatic failures in Korea, Vietnam, Iran, Granada, and many others. Getting the military services to work together has been a recurring battle that has had to be addressed time and again. But despite the problems, we understood that the nature of conflict required us to develop and support key capabilities and institutions and, over time, devote the necessary resources, people and money, and get enough things right while maintaining the ability to recover from mistakes along the way. I suggest this is our task today.

To this end, the Department of Defense will soon award a contract to an independent, non-partisan, non-profit group to produce a study that in effect tries to answer the question I posed at Kansas State. If we were to rewrite the National Security Act of 1947 for the 21st century, what would it look like? What new institutions, arrangements, and authorities would it create? I look forward to seeing the result, which perhaps might form the basis of legislation or at least debate in the next administration.

In closing, I have observed that repeatedly over the last century, Americans averted their eyes in the belief that events in remote places around the world need not engage this country. How could an assassination of an Austrian archduke in unknown Bosnia-Herzegovina affect us, or the annexation of a little patch of ground called Sudetenland, or a French defeat in a place called Dien Bien Phu, or the return of an obscure cleric to Tehran, or the radicalization of a Saudi construction tycoon's son?

What seems to work best in world affairs, historian Donald Kagan wrote in his book, "On the Origins of War," is that the possession by those states who wish to preserve the peace of the preponderant power and of the will to accept the burdens and responsibilities required to achieve that purpose. In an address at Harvard in 1943, Winston Churchill said, "the price of greatness is responsibility." The people of the United States cannot escape world responsibility. Our country has now, for many decades, taken upon itself great burdens and great responsibilities, all in an effort to defeat despotism in its many forms and to preserve the peace.

Today, across the globe, there are more people than ever seeking economic and political freedom, seeking hope, even as repressive regimes and mass murderers sow chaos in their midst. For all of those brave men and women struggling for a better life, there is and must be no stronger ally or advocate than the United States of America. Our responsibilities to them and to the world in the final analysis are not a burden on the people or the soul of this nation. They are, rather, a blessing. Thank you.

## Senate Armed Services Committee Opening Remarks

*Washington, D.C., Wednesday, February 06, 2008*

THANK you, Mr. Chairman. Mr. Chairman, members of the committee, it is a pleasure to be here for my second and last posture statement.

Let me first thank you for your continued support for our military these many years. I appreciate the opportunity today to discuss the president's Fiscal Year 2009 Defense budget request.

Before getting into the components of the request, I thought it might be useful to consider it quickly in light of the current strategic landscape, a landscape still being shaped by forces unleashed by the end of the Cold War nearly two decades ago. In recent years old hatreds and conflicts have combined with new threats and forces of instability—challenges made more dangerous and prolific by modern technology. Among them: terrorism, extremism and violent jihadism; ethnic, tribal and sectarian conflict; proliferation of dangerous weapons and materials; failed and failing states; nations discontented with their role in the international order; and rising and resurgent powers whose future paths are uncertain.

In light of this strategic environment, we must make the choices and investments necessary to protect the security, prosperity and freedom of Americans for the next generation.

The investment in Defense spending being presented today is $515.4 billion, or about 4 percent of our gross domestic product when combined with war costs. This compares to spending levels of 14 percent of gross domestic product during the Korean War and 9 percent during Vietnam. Our FY 2009 request is a 7.5 percent increase, or $35.9 billion, over last year's enacted level. When accounting for inflation, this translates into a real increase of about 5-1/2 percent.

The difference consists of four main categories which are outlined in more detail in my submitted station—statement.

Overall, the budget includes $183.8 billion for overall strategic modernization, including $104 billion for procurement, to sustain our nation's technological advantage over current and future adversaries; $158.3 billion for operations, readiness and support to maintain a skilled and agile fighting force; $149.4 billion to enhance quality of life for our men and women in uniform by providing the pay, benefits, health care and other services earned by our all-volunteer force; and $20.5 billion to increase ground capabilities by growing the Army and the Marine Corps.

This budget includes new funding for critical ongoing initiatives such as global train-and-equip to build the security capacity of partner nations, security and stabilization assistance, foreign language capabilities, and the new Africa Command.

In summary, this request provides the resources needed to respond to current threats while preparing for a range of conventional and irregular challenges that our nation may face in the years ahead.

In addition to the 515 billion—$515.4 billion base budget, our request includes $70 billion in emergency bridge funding that would cover war costs in the next—into the next calendar year. A more detailed request will be submitted later this year, when the department has a better picture of what level of funding will be needed.

The 2007 National Defense Authorization Act, as you have pointed out, requires the Department of Defense to provide an estimate of costs for the global war on terror. We would like to be responsive to the request. Indeed, I was responsive to a similar request last year. Some have alleged that the administration has taken this position in order to somehow hide the true costs of the war. Nothing could be further from the truth. The department has been very open about what we know about our costs, as well as what we don't know. So the challenge we face is that a realistic or meaningful estimate requires answers to questions that we don't yet know, such as when and if the department will receive the requested $102 billion balance of the FY 2008 supplemental war request, and for how much; and what, if any, adjustment to troop levels in Iraq will result from the upcoming recommendations of General Petraeus, Central Command and the Joint Chiefs of Staff. We should also keep in mind that nearly three- quarters of the FY 2009 supplemental request will likely be spent in the next administration, thus making it even more difficult to make an accurate projection.

I've worked hard during my time in this job to be responsive and transparent to this committee and to the Congress. Nothing has changed. But while I would like to be in a position to give you a realistic estimate of what the department will need for the FY 2009 supplemental funds, I simply cannot at this point.

There are too many significant variables in play.

I can give you a number—I will give you a number if you wish, but I will tell you that the number will inevitably be wrong, and perhaps significantly so. So I will be giving you precision without accuracy. As I mentioned earlier, Congress has yet to appropriate the remaining balance of the FY 2008 war funding request, $102.5 billion. The delay is degrading our ability to operate and sustain the force at home and in the theater, and is making it difficult to manage this department in a way that is fiscally sound. The Department of Defense, as I've said, is like the world's biggest supertanker. It cannot turn on a dime and cannot be steered like a skiff. I urge approval of the FY 2008 request as quickly as possible.

Finally, I would like to thank the members of this committee for all you have done to support our troops as well as their families. I thank you specifically for your attention to and support of efforts to improve the treatment of wounded warriors over the past year. In visits to the combat theaters, in military hospitals, and in bases and posts at home and around the world, I continue to be amazed by the decency, resilience, and courage of our troops. Through the support of the Congress and our nation, these young men and women will prevail in the current conflicts and be prepared to confront the threats that they, their children, and our nation may face in the future.

Thank you, Mr. Chairman.

# House Armed Services Committee Opening Remarks

*Washington, D.C., Wednesday, February 06, 2008*

(OFF mike)—it's a pleasure to be here for my second and last posture statement. Let me thank you, first of all, for your continued support of our military for these many years. And I appreciate the opportunity to discuss the president's fiscal year 2009 defense budget.

Before getting into the components of the request, I thought it might be useful to consider, in light of the current strategic landscape, a landscape still being shaped by forces unleashed by the end of the Cold War nearly two decades ago. In recent years, old hatreds and conflicts have combined with new threats and forces and instability, challenges made more dangerous and prolific by modern technology.

Among them terrorism, extremism, violent jihadism, ethnic tribal and sectarian conflict, proliferation of dangerous weapons of—and materials, failed and failing states, nations discontented with their role in the international order and rising and resurgent powers whose future paths are uncertain.

In light of this strategic environment—a complex strategic environment, we must make the choices and invest in what's necessary to protect the security, prosperity and freedom of Americans for the next generation. The investment in defense spending being presented today is $515.4 billion, or about 4 percent of our gross domestic product when combined with war costs. This compares to spending levels of 14 percent of gross domestic product during the Korean War and 9 percent during Vietnam. Our Fiscal Year 2009 request is a 7.5 percent increase or $35.9 billion over last year's enacted level. When accounting for inflation, this translates into a real increase of about 5-and-a-half percent.

The difference consists of four main categories which are outlined in more detail in my submitted statement. Overall, the budget includes $183.8 billion for overall strategic modernization, to include $104 billion for procurement to sustain our nation's technological advantage over current and future adversaries; $158.3 billion for operations, readiness and support to maintain a skilled and agile fighting force; $149.4 billion to enhance the quality of life by providing the pay, benefits, health care and other services earned by our all-volunteer force, and $20.5 billion to increase ground capabilities by growing the Army and Marine Corps. This budget includes new funding for critical ongoing initiatives such as Global Train and Equip to build the security capacity of partner nations, security and stabilization assistance, foreign language capabilities, a real increase in science and technology basic research and the new Africa Command.

In summary, this request provides the resources needed to respond to current threats while preparing for a range of conventional and irregular challenges that our nation may face in the years ahead. In addition to the $515.4 billion base budget, our request includes $70 billion in emergency bridge funding that would cover war costs into the next calendar year. A more detailed request will be submitted when the department has a better picture of what level of funding will be needed. The 2007 National Defense Authorization Act requires the Department of Defense to provide an estimate of costs for the Global War on Terror. We would like to be responsive to this request. Indeed, I was responsive to a similar request last year. Some have alleged the administration has taken this position in order, somehow, to hide the true costs of the war. Nothing could be further from the truth. The department has been very open about what we know about our costs as well as what we don't know. So the challenge we face is that a realistic or meaningful estimate requires answers to questions that we don't yet know, such as when and if the department will receive the requested $102 billion balance of the FY08 supplemental war request and for how much, and what—if any—adjustments to troop levels in Iraq will result from the upcoming recommendations of General Petraeus. We should also keep in mind that three-quarters of the FY09 supplemental request will likely be spent in the next administration, thus making it even more difficult to make an accurate projection.

I've worked very hard during my time in this position to be responsive and transmit—transparent to this committee and to the Congress. Nothing has changed. But while I would like to be in a position to give you a realistic estimate of what the department will need for an FY 2009 supplemental and will do so at the earliest possible time, I simply can't at this point. There are too many significant variables in play. I can give you a number, but that number would inevitably be wrong, perhaps significantly so. In short, precision without accuracy.

As I mentioned earlier, Congress has yet to appropriate the remaining balance of the FY08 War Funding Request—$102.5 billion. This delay is degrading our ability to operate and sustain the force at home and in the theater, and is making it difficult to manage the department in a way that is fiscally sound. The Department of Defense is like the world's biggest supertanker, it cannot turn on a dime and it cannot be steered like a skiff. I urge approval of the 2008 GWAT request (ph) as quickly as possible.

Finally, I would like to thank the members of this committee for all you have done to support our troops as well as their families. I also thank you for your attention to and your support of efforts to improve the treatment of our wounded warriors over the past year. In visits to the combat theaters, military hospitals and in bases and posts at home, and around the world, I continue to be amazed by the decency, resilience and courage of our troops. Through support of the Congress and our nation, these young men and women will prevail in the current conflicts and be prepared to confront the threats that they, their children and our nation may face in the future.

Thank you, Mr. Chairman.

# Appendix I

# Dr. Robert M. Gates

Dr. Robert M. Gates was sworn in on December 18, 2006, as the 22nd Secretary of Defense. Before entering his present post, Secretary Gates was the President of Texas A&M University, the nation's seventh largest university.

Prior to assuming the presidency of Texas A&M on August 1, 2002, he served as Interim Dean of the George Bush School of Government and Public Service at Texas A&M from 1999 to 2001.

Secretary Gates served as Director of Central Intelligence from 1991 until 1993. Secretary Gates is the only career officer in CIA's history to rise from entry-level employee to Director He served as Deputy Director of Central Intelligence from 1986 until 1989 and as Assistant to the President and Deputy National Security Adviser at the White House from January 20, 1989, until November 6, 1991, for President George H.W. Bush.

Secretary Gates joined the Central Intelligence Agency in 1966 and spent nearly 27 years as an intelligence professional, serving six presidents. During that period, he spent nearly nine years at the National Security Council, The White House, serving four presidents of both political parties.

Secretary Gates has been awarded the National Security Medal, the Presidential Citizens Medal, has twice received the National Intelligence Distinguished Service Medal, and has three times received CIA's highest award, the Distinguished Intelligence Medal.

He is the author of the memoir, From the Shadows: The Ultimate Insiders Story of Five Presidents and How They Won the Cold War, published in 1996.

Until becoming Secretary of Defense, Dr. Gates served as Chairman of the Independent Trustees of The Fidelity Funds, the nation's largest mutual fund company, and on the board of directors of NACCO Industries, Inc., Brinker International, Inc. and Parker Drilling Company, Inc.

Dr. Gates has also served on the Board of Directors and Executive Committee of the American Council on Education, the Board of Directors of the National Association of State Universities and Land-Grant Colleges, and the National Executive Board of the Boy Scouts of America. He has also been President of the National Eagle Scout Association.

A native of Kansas, Secretary Gates received his bachelor's degree from the College of William and Mary, his master's degree in history from Indiana University, and his doctorate in Russian and Soviet history from Georgetown University.

In 1967 he was commissioned a second lieutenant in the U.S. Air Force and served for a year as an intelligence officer at Whiteman Air Force Base in Missouri.

Dr. Gates and his wife Becky have two adult children.

# Appendix II

# Responsibilities of the Office of the Secretary of Defense

The Office of the Secretary of Defense (OSD) is the principal staff element used by the Secretary and Deputy Secretary of Defense to exercise authority, direction, and control over the Department of Defense. The mission of OSD as an organizational entity, in coordination with other elements of DoD, is as follows:

- Develop and promulgate policies in support of United States national security objectives.
- Provide oversight to assure the effective allocation and efficient management of resources consistent with Secretary of Defense approved plans and programs.
- Develop appropriate evaluation mechanisms to provide effective supervision of policy implementation and program execution at all levels of the Department.
- Provide the focal point for departmental participation in the United States security community and other Government activities.

In addition, each OSD principal staff official, in his/her respective area of functional assignment, is responsible for performing the following:

- Conduct analyses, develop policies, provide advice, make recommendations, and issue guidance on Defense plans and programs.
- Develop systems and standards for the administration and management of approved plans and programs.
- Initiate programs, actions, and taskings to ensure adherence to DoD policies and national security objectives, and to ensure that programs are designed to accommodate operational requirements.
- Review and evaluate programs for carrying out approved policies and standards.
- Inform appropriate organizations and personnel of new and significant trends or initiatives in assigned areas of functional responsibilities.
- Review proposed resource programs, formulate budget estimates, recommend resource allocations, and monitor the implementation of approved programs.
- Participate in those planning, programming, and budgeting activities, which relate to assigned areas of functional responsibilities.
- Review and evaluate recommendations on requirements and priorities.

- Promote coordination, cooperation, and mutual understanding within the Department of Defense and between DoD and other Federal agencies and the civilian community.
- Serve on boards, committees, and other groups pertaining to assigned functional areas, and represent the Secretary of Defense on matters outside the Department of Defense.
- Develop information and data, prepare reports, and/or testimony for presentations to Congressional Committees or in response to Congressional inquiries.
- Represent the DoD with Congressional Committees or individual Members of the Congress.
- Perform such other duties as the Secretary of Defense may from time to time prescribe.

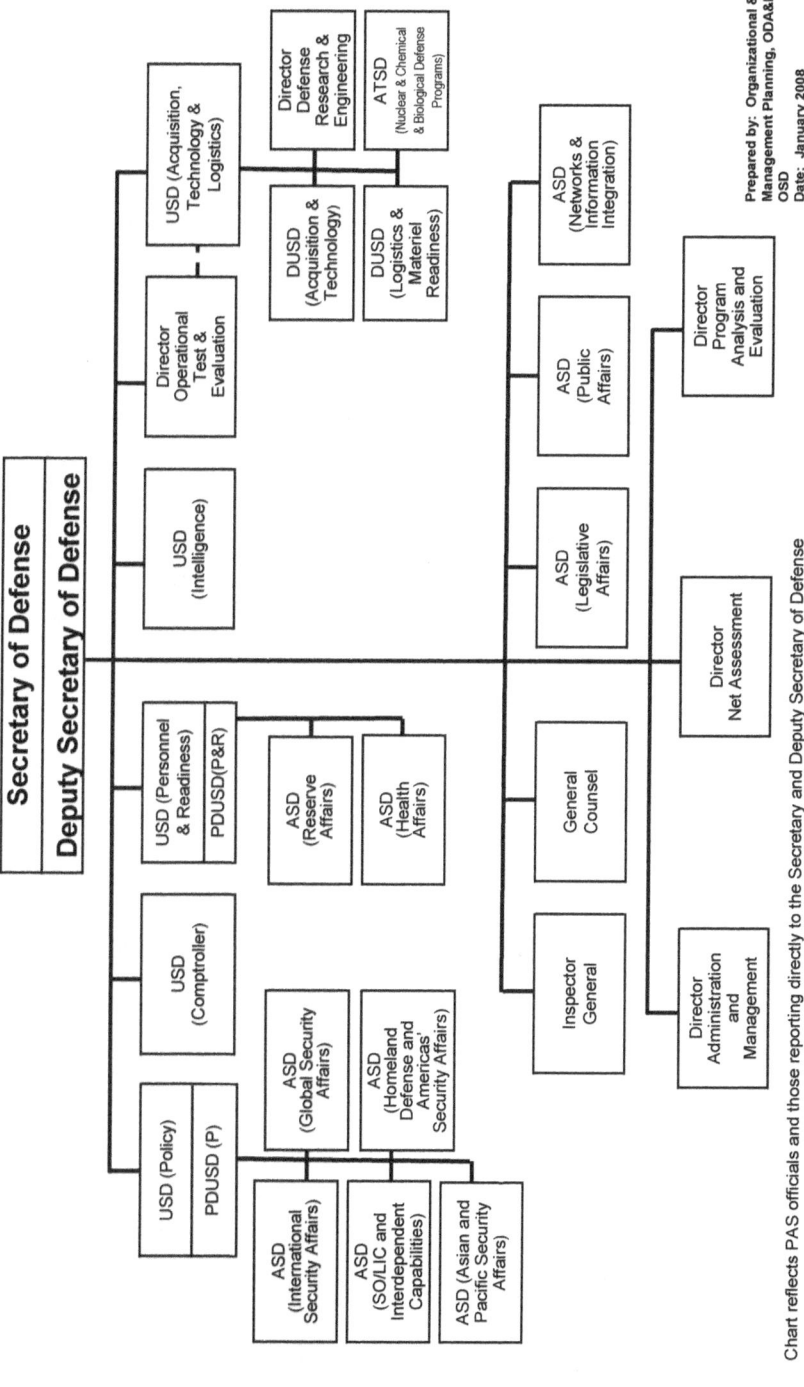

# Appendix IV

## Munich Conference on Security Policy

*Munich, Germany, Sunday, February 10, 2008*

(This speech was made available after the book had already been readied for print. However, due to possible new initiatives included in this speech, we were able to include it as an appendix. Please note that this speech is not included in the index.)

THANK you, Horst. I would also like to thank the people of Munich for once again allowing us to gather in this beautiful city.

I am glad to see many of my colleagues here, as well as many of the delegations that were with us in Vilnius for the NATO ministerial. As I said in Vilnius – three weeks ago I accomplished a key goal I have been pursuing for the past year: through the good offices of the Los Angeles Times, I finally brought unity to NATO – though not as I wished.

It is an honor to be invited to speak here for a second, and last, year as U.S. Secretary of Defense.

Vilnius was my fourth NATO ministerial since taking this post, but my first in a nation that had been part of the former Soviet Union. Lithuania was one of the first nations to be swallowed by the Soviets, and the first republic to declare its independence as Baltic push came to Soviet shove. It is now a proud member of NATO, and the leader of a Provincial Reconstruction Team in Afghanistan.

For the transatlantic alliance, the period in which Lithuania and other captive nations gained their independence was a time of reflection. Not only were we pondering enlargement to secure the wave of democracy sweeping across Eastern Europe, but NATO was also pondering the very concept of collective self-defense in a post- Cold War world.

We saw this in 1991, when NATO issued its first Strategic Concept. This document recognized that a "single massive and global threat ha[d] given way to diverse and multi-directional risks" – challenges such as weapons proliferation; disruption of the flow of vital resources; ethnic conflict; and terrorism. Overcoming these threats, the document stated, would require a "broad approach to security," with political, economic, and social elements.

From the perspective of one who played a role in that effort to redirect NATO 17 years ago, today I would like to discuss a subject that embodies the security challenges that have emerged since that time, and correspondingly, the capabilities we need, in this new era.

That subject is, not surprisingly, Afghanistan. After six years of war, at a time when many sense frustration, impatience, or even exhaustion with this mission, I believe it is valuable to step back and take stock of Afghanistan:

- First, within the context of the long-standing purpose of the Alliance, and how it relates to the threats of a post Cold War world;
- Second, with regard to NATO's vision of becoming a transformed, multifaceted, expeditionary force – and how we have evolved in accordance with that vision; and
- Finally, to recapitulate to the people of Europe the importance of the Afghanistan mission and its relationship to the wider terrorist threat.

There is little doubt that the mission in Afghanistan is unprecedented. It is, in fact, NATO's first ground war and it is dramatically different than anything NATO has done before. However, on a conceptual level, I believe it falls squarely within the traditional bounds of the Alliance's core purpose: to defend the security interests and values of the transatlantic community.

During the 1990s, even as we tried to predict what form the threats of the 21st century would take, Afghanistan was, in reality becoming exactly what we were discussing in theory. Subsequent events during the ensuing years have shown that:

- Instability and conflict abroad have the potential to spread and strike directly at the hearts of our nations;
- New technology and communications connect criminal and terrorist networks far and wide, and allow local problems to become regional and even global;
- Economic, social, and humanitarian problems caused by massive immigration flows radiate outward with little regard for national borders;
- A nexus between narcotics and terrorists increases the resources available to extremists in the region, while increasing the drug flow to European streets; and
- The presence of safe havens, combined with a lack of development and governance, allow Islamic extremists to turn a poisonous ideology into a global movement.

More than five years ago in Prague, in the wake of the September 11th attacks, our nations set out to transform NATO into an expeditionary force capable of dealing with threats of this type – capable of helping other nations help themselves to avoid Afghanistan's fate. At the time, I imagine many were unsure of what, exactly, this would look like – what new structures, training, funding, mindsets, and manpower would be needed. Since then, however, we have applied our vision on the ground in Afghanistan.

Today:

- Nearly 50,000 troops from some 40 allies and partner nations serve under NATO command, thousands of miles from the Alliance's traditional borders;
- Growing numbers of reconstruction and security training teams are making a real difference in the lives of the Afghan people; and
- NATO's offensive and counterinsurgency operations in the South have dislodged the Taliban from their strongholds and reduced their ability to launch large scale or coordinated attacks.

Due to NATO's efforts, as Minister Jung pointed out yesterday, Afghanistan has made substantial progress in health care, education, and the economy – bettering the lives of millions of its citizens.

Through the Afghan mission, we have developed a much more sophisticated understanding of what capabilities we need as an Alliance and what shortcomings must be addressed.

Since the Riga summit, there has been much focus on whether all allies are meeting their commitments and carrying their share of the burden. I have had a few things to say about that myself. In truth, virtually all allies are fulfilling the individual commitments they have made. The problem is that the Alliance as a whole has not fulfilled its broader commitment from Riga to meet the force requirements of the commander in the field.

As we think about how to satisfy those requirements, we should look more creatively at other ways to ensure that all allies can contribute more to this mission – and share this burden. But we must not – we cannot – become a two-tiered Alliance of those who are willing to fight and those who are not. Such a development, with all its implications for collective security, would effectively destroy the Alliance.

As many of you know, a Strategic Vision document is being drafted that will assess NATO's and our partners' achievements in Afghanistan, and will produce a set of realistic goals and a roadmap to meet them over the next three to five years. We continue urgently to need a senior civilian – a European in my view – to coordinate all non-military international assistance to the Afghan government and people. The lack of such coordination is seriously hampering our efforts to help the Afghans build a free and secure country.

The really hard question the Alliance faces is whether the whole of our effort is adding up to less than the sum of its parts, and, if that is the case, what we should do to reverse that equation.

As an Alliance, we must be willing to discard some of the bureaucratic hurdles that have accumulated over the years and hinder our progress in Afghanistan. This means more willingness to think and act differently – and quickly. To pass initiatives such as the NATO Commander's Emergency Response Fund. This tool has proven itself elsewhere, but will, for NATO, require a more flexible approach to budgeting and funding.

Additionally, it is clear that we need a common set of training standards for every one going to Afghanistan – whether they are combat troops conducting counterinsurgency operations; civilians working in Provincial Reconstruction Teams; or members of operational mentoring and liaison training teams. Unless we are all on the same page – unless our efforts are tied together and unified by similar tactics, training, and goals – then the whole of our efforts will indeed be less than the sum of the parts.

I also worry that there is a developing theology about a clear-cut division of labor between civilian and military matters – one that sometimes plays out in debates over the respective roles of the European Union and NATO, and even among the NATO allies. In many respects, this conversation echoes one that has taken place – and still is – in the United States within the civilian and military agencies of the U.S. government as a result of the Afghanistan and Iraq campaigns.

For the United States, the lessons we have learned these past six years – and in many cases re-learned – have not been easy ones. We have stumbled along the way, and we are still learning. Now, in Iraq, we are applying a comprehensive strategy that emphasizes the security of the local population – those who will ultimately take control of their own security – and brings to bear in the same place and very often at the same time civilian resources for economic and political development.

We have learned that war in the 21st century does not have stark divisions between civilian and military components. It is a continuous scale that slides from combat operations to economic development, governance and reconstruction – frequently all at the same time.

The Alliance must put aside any theology that attempts clearly to divide civilian and military operations. It is unrealistic. We must live in the real world. As we noted as far back as 1991, in the real world, security has economic, political, and social dimensions. And vice versa. The E.U. and NATO need to find ways to work together better, to share certain roles – neither excluding NATO from civilian operations nor barring the E.U. from military missions. In short, I agree entirely with Secretary General de

Hoop Scheffer and Minister Morin's comments yesterday that there must be a "complimentarity" between the E.U. and NATO.

At the same time, in NATO, some allies ought not to have the luxury of opting only for stability and civilian operations, thus forcing other Allies to bear a disproportionate share of the fighting and the dying.

Overall, the last few years have seen a dramatic evolution in NATO's thinking and in its posture. With all the new capabilities we have forged in the heat of battle – and with new attitudes – we are seeing what it means to be expeditionary. What is required to spread stability beyond our borders. We must now commit ourselves to institutionalize what we have learned and to complete our transformation.

Just as we must be realistic about the nature and complexity of the struggle in Afghanistan, so too must we be realistic about politics in our various countries. NATO, after all, is an alliance whose constituent governments all answer to their citizens.

My colleagues in Vilnius and those in this room certainly understand the serious threat we face in Afghanistan. But I am concerned that many people on this continent may not comprehend the magnitude of the direct threat to European security. For the United States, September 11th was a galvanizing event – one that opened the American public's eyes to dangers from distant lands. It was especially poignant since our government had been heavily involved in Afghanistan in the 1980s, only to make the grievous error – of which I was at least partly responsible – of abandoning a destitute and war-torn nation after the last Soviet soldier crossed the Termez bridge.

While nearly all the Alliance governments appreciate the importance of the Afghanistan mission, European public support for it is weak. Many Europeans question the relevance of our actions and doubt whether the mission is worth the lives of their sons and daughters. As a result, many want to remove their troops. The reality of fragile coalition governments makes it difficult to take risks. And communicating the seriousness of the threat posed by Islamic extremism in Afghanistan, the Middle East, Europe, and globally remains a steep challenge.

As opinion leaders and government officials, we are the ones who must make the case publicly and persistently.

So now I would like to add my voice to those of many allied leaders on the continent and speak directly to the people of Europe: The threat posed by violent Islamic extremism is real – and it is not going away. You know all too well about the attacks in Madrid and London. But there have also been multiple smaller attacks in Istanbul, Amsterdam, Paris, and Glasgow, among others. Numerous cells and plots have been disrupted in recent years as well – many of them seeking large-scale death and destruction, such as:

- A complex plot to down multiple airliners over the Atlantic that could have killed hundreds or thousands;
- A plot to use ricin and release cyanide in the London Underground;
- A separate plan for a chemical attack in the Paris metro;
- Plots in Belgium, England, and Germany involving car bombs that could have killed hundreds;
- Homemade bombs targeting commuter and high-speed trains in Spain and Germany;
- Individuals arrested in Bosnia with explosives, a suicide belt, and an instructional propaganda video;
- Two plots in Denmark involving explosives, fertilizer, and a bomb-making video; and
- Just in the last few weeks, Spanish authorities arrested 14 Islamic extremists in Barcelona suspected of planning suicide attacks against public transport systems in Spain, Portugal, France, Germany, and Britain.

Imagine, for a moment, if some or all of these attacks had come to pass. Imagine if Islamic terrorists had managed to strike your capitals on the same scale as they struck in New York. Imagine if they had laid their hands on weapons and materials with even greater destructive capability – weapons of the sort all too easily accessible in the world today. We forget at our peril that the ambition of Islamic extremists is limited only by opportunity.

We should also remember that terrorist cells in Europe are not purely homegrown or unconnected to events far away – or simply a matter of domestic law and order. Some are funded from abroad. Some hate all western democracies, not just the United States. Many who have been arrested have had direct connections to Al Qaeda. Some have met with top leaders or attended training camps abroad. Some are connected to Al Qaeda in Iraq. In the most recent case, the Barcelona cell appears to have ties to a terrorist training network run by Baitullah Mehsud, a Pakistan-based extremist commander affiliated with the Taliban and Al Qaeda – who we believe was responsible for the assassination of Benazir Bhutto.

What unites them is that they are all followers of the same movement – a movement that is no longer tethered to any strict hierarchy but one that has become an independent force of its own. Capable of animating a corps of devoted followers without direct contact. And capable of inspiring violence without direct orders.

It is an ideological movement that has, over the years, been methodically built on the illusion of success. After all, about the only thing they have accomplished recently is the death of thousands of innocent Muslims while trying to create discord across the Middle East. So far they have failed. But they have twisted this reality into an aura of success in many parts of the world. It raises the question: What would happen if the false success they proclaim became real success? If they triumphed in Iraq or Afghanistan, or managed to topple the government of Pakistan? Or a major Middle Eastern government?

Aside from the chaos that would instantly be sown in the region, success there would beget success on many other fronts as the cancer metastasized further and more rapidly than it already has. Many more followers could join their ranks, both in the region and in susceptible populations across the globe. With safe havens in the Middle East, and new tactics honed on the battlefield and transmitted via the Internet, violence and terrorism worldwide could surge.

I am not indulging in scare tactics. Nor am I exaggerating either the threat or inflating the consequences of a victory for the extremists. Nor am I saying that the extremists are ten feet tall. The task before us is to fracture and destroy this movement in its infancy – to permanently reduce its ability to strike globally and catastrophically, while deflating its ideology. Our best opportunity as an alliance to do this is in Afghanistan. Just as the hollowness of Communism was laid bare with the collapse of the Soviet Union, so too would success in Afghanistan, as well as in Iraq, strike a decisive blow against what some commentators have called Al Qaeda-ism.

This is a steep challenge. But the events of the last year have proven one thing above all else: If we are willing to stand together, we can prevail. It will not be quick, and it will not be easy – but it can be done.

In the years ahead, the credibility of NATO, and indeed the viability of the Euro-Atlantic security project itself, will depend on how we perform now. Other actors in the global arena – Hezbollah, Iran and others – are watching what we say and what we do, and making choices about their future course.

Everyone knows that in 2009 the United States will have a new administration. And this time, next year, you will be hearing from a new Secretary of Defense.

But regardless of which party is in power, regardless who stands at this podium, the threats we face now and in the future are real. They will not go away. Overcoming them will require unity between opposition parties and across various governments, and uncommon purpose within the Alliance and with other friends and partners.

I began my remarks with a bit of history about NATO in the 1990s. I would like to close with a few words about the dawn of the transatlantic Alliance.

From our present-day vantage point, victory in the Cold War now seems almost preordained. But as we prepare to celebrate NATO's 60th anniversary next year, it is useful to recall that 60 years ago this year, in 1948, the year of the Berlin airlift, few people would have been all that optimistic about the future of Europe, or the prospect of a Western alliance. The Continent was devastated, its economy in shambles. The United States was debating the European recovery program – known as the Marshall Plan – and faced a resurgent isolationism. Europe was under siege – with pressure from communism being felt in Germany, France, Finland, Norway, Italy, Czechoslovakia, and Greece.

In January of that year, Ernest Bevin, the British foreign secretary, went before parliament to discuss the Soviet Union and other threats to the United Kingdom. Between all the "kindred souls of the West," he said, "there should be an effective understanding bound together by common ideals for which the Western Powers have twice in one generation shed their blood."

Less than two months later, President Harry Truman stood in the United States Congress and echoed that sentiment. He said: "The time has come when the free men and women of the world must face the threat to their liberty squarely and courageously . . . Unity of purpose, unity of effort, and unity of spirit are essential to accomplish the task before us."

That unity held for decades through ups and downs. It held despite divisions and discord, stresses and strains, and through several crises where another war in Europe loomed. Alexis de Tocqueville once warned that democracies, when it comes to foreign affairs, were ill-suited to pursue a "great undertaking" and "follow it [through] with determination." But the democracies of the West did just that – for more than 40 years. And they can do so once more today.

We must find the resolve to confront together a new set of challenges. So that, many years from now, our children and their children will look back on this period as a time when we recommitted ourselves to the common ideals that bind us together. A time when we again faced a threat to peace and to our liberty squarely and courageously. A time when we again shed blood and helped war devastated people nourish the seeds of freedom and create peaceful, productive societies. That mission drew us together in 1948 and keeps us together today.

Many years from now, perhaps future generations will look back on this period and say, "victory seemed almost preordained."

Thank you.

# Index

1st Brigade of the First Armored Division  162

## A

A-10  47
Abrams, Creighton - General  217
Abshire, Dave - CSIS co-founder  239
Abu Sayyaf  114
Adams, Abigail  97, 98, 101, 189, 190
Adams, John - President  98, 108, 141, 170, 190
Adams, John Quincy - President  97, 98, 101, 140, 189
Adams, Thomas Boylston  98, 190
Aeronautics  106
Aerospace Industries Association  91
Afghanistan  8, 19, 22, 23, 24, 27, 28, 29, 30, 31, 32, 33, 35, 36, 37, 39, 40, 44, 47, 53, 59, 62, 63, 71, 74, 78, 80, 82, 85, 87, 89, 90, 96, 111, 112, 114, 115, 116, 124, 129, 130, 131, 133, 136, 139, 140, 141, 142, 144, 145, 147, 151, 152, 153, 155, 159, 161, 167, 171, 173, 174, 175, 177, 180, 181, 183, 186, 195, 202, 204, 207, 213, 216, 217, 218, 219, 226, 229, 230, 231, 232, 235, 240, 241, 243, 244
Afghan National Security Forces (ANSF)  23, 33, 63, 89
Africa  112, 114, 140, 242, 244, 247, 250
Africa Command  244, 247, 250
African Americans  233, 234
Agency for International Development (AID)  43, 80, 132, 162, 215, 216, 219, 241, 244
Air Force  20, 21, 39, 43, 45, 47, 48, 60, 61, 74, 80, 86, 87, 105, 106, 107, 109, 131, 136, 146, 147, 148, 155, 161, 176, 179, 199, 202, 203, 204, 211, 254
Air Force Academy  105, 106
Air Force museum  48
Air Guard  38
Airmen  105, 106, 146, 147, 204
airpower  106, 147
Al Anbar  163
Alaska  118
Albright, Madeleine - Secretary of State/Congressional Liason  172
Alcibiades  104
Algeria  27
Algiers  55, 71, 172, 225
Allard - Senator  45
Allen, Admiral  123
Allied victory  213
All-Volunteer Force  21, 61, 88
Al Qaeda  23, 27, 33, 71, 114, 116, 124, 131, 132, 144, 162, 163, 174, 223, 225, 231
al-Taei, Ahmed - Specialist  150
A Man Called Intrepid  213
Ambrose, Stephen  66
American Chamber of Commerce of Cairo  69
American Council on Education  253
American democracy  35, 95, 233
American Revolution  97, 101, 141, 158
American-Turkish Council  52
America Supports You  44, 78, 237, 238
amphibious assault  129
Anbar Awakening  163
Andrews Air Force Base  131
Angola  82, 136
Annapolis  99, 134, 135, 228
anti-American  54
anti-satellite  20, 29, 60, 86, 177
anti-Semitism  170
Apaches  204
Appropriations  31, 58, 84, 151
Appropriations Bill  151
Arab  71, 72, 161, 172, 175, 220, 225
Arab Americans  161
Arabic  132, 163
Arab nationalism  175
Ardennes  121

Arlington  127, 134, 150, 169
Armstrong, Anne - Ambassador  239
Army  10, 11, 12, 13, 15, 16, 19, 20, 21, 23, 32, 36, 37, 38, 50, 59, 60, 61, 62, 65, 66, 67, 68, 80, 81, 85, 86, 88, 89, 97, 106, 124, 125, 130, 152, 157, 158, 159, 160, 161, 162, 163, 167, 180, 206, 207, 208, 209, 214, 216, 218, 221, 234, 236, 242, 243, 247, 250
Army Chief of Staff  65, 124
Army Guard  36
Army-Navy football game  50
Army Spouse Employment Partnership program  208
Arnold, "Hap" - General  105, 147
ASEAN  117
Asia  26, 28, 69, 82, 111, 112, 113, 114, 115, 116, 117, 118, 124, 129, 147, 191, 192, 193, 194, 195, 212
Asian security  111, 116, 193
Asia Pacific region  111
Association of the U.S. Army (AUSA)  157, 158, 208
asymmetric war  131, 136
Ataturk, Kemal  54
Atlantic Alliance  27, 55, 143
Atlantic partnership  26, 183
Australia  28, 29, 50, 115, 120, 191, 192, 193
Australia's Reconstruction Task Force  115
Austria  164
Avery, Brad  237
avian influenza  113, 194
Aviation Bridging Force  183

## B

B-2  106, 202, 205
Baghdad  9, 14, 17, 24, 33, 55, 63, 71, 72, 207, 218, 223, 228, 235
Bagram  96
Bahrain  222
Baker-Hamilton Commission  93, 139
Bali  112, 193

Bali bombings  193
Balkans  30, 142, 145, 161, 167, 212, 216, 243
ballistic-missile facility  45
ballistic missiles  20, 60, 86, 112, 179, 194, 226
Bancroft Hall  99
Barak, Ehud - Prime Minister/Chief of Staff (Israel)  171
barracks  21, 40, 61, 88, 131
Barrow, Isaac  213
Base Budget  19, 59, 85
Base Community Council  202, 204
Basic Allowance for Housing  21, 61, 88
Becken, Forest - Scout Master  200
Beirut  72, 131, 228
Belousov - General  164
Berlin  26, 82, 117, 147
Berlin Wall  82, 117
Bethesda  153, 158
Bilateral Air Defense Initiative  227
Black Hawk  125
Bloody Omaha  120
Bloomington  202
Blue Angels  99
Blum, General  35, 38
Board of Directors  253
Bong, Dick  105, 147
Bonn  25
Bono  237
Bosnia-Herzegovina  220, 245
Bowen, Ray - President of A&M  185
Boyda -Congresswoman  210
Boy Scouts  199, 200, 201, 253
Brantley, John - Private  35
Brazos  188
Brezhnev - General Secretary (USSR)  164
Brezinsky, Zbigniew - National Security Adviser  55, 71, 171, 172
Bright Star exercise  69
Brinker International, Inc.  253
Brinkley, David  52, 128, 158

Brinkley, David - Newsman  52, 128, 158
Brooke  158, 208
Brooke Army Medical Center  208
Brookhiser, Richard  95
Brown, Doug - General  124
Brown, Harold - Secetary of Defense  56, 124, 125
Buckley, William - CIA Station Chief, Beirut  131
Budget, Defense  18, 58, 84
Bunning - Senator  105
Burr, Aaron  187
Bush, Barbara - First Lady  187
Bush, George H.W. - President  184, 188, 211, 253
Bush, George W. - President  27, 41, 55, 112, 113, 143, 183, 184, 185, 186, 189, 192, 194, 207, 222, 223, 228, 241, 243
Bush School  185, 186, 253
Byzantine  54

## C

Cairo  69, 224
California  118, 161, 204
Camp David  69, 171
Camp David Accords  69
Canada  28, 120
Cape Canaveral  92
Capitol  94, 230, 241
Capitol Hill  230, 241
Caribbean  179
Carter Administration  56, 132
Carter Doctrine  222
Carter, Jimmy - President  211, 222
Cartwright, Hoss  177
Casey, George - General  13, 14, 16, 17, 67, 68, 123, 159
Catholic  212
cavalry  106, 147
CENTCOM  123
Center for Strategic and International Studies (CSIS)  239

Central Asia  26, 28, 111, 113, 115, 192, 193, 194
Central Europe  140, 143
Central Intelligence Agency  211, 225, 253. *See also* CIA
Central Texas communities  208
Cessna  203
Chancellor's Bungalow  25
Change of Responsibility Ceremony  65
Chao, Elaine - Secretary of Labor  197
Cheney, Dick - Vice President/Secretary of Defense  131, 135
cherry trees  196
Chiarelli - General  207
Chief of Naval Operations  158, 220
Chilcoat, Dick - Lieutenant General  185
Chilton, Kevin "Chili" Chilly - General  49, 176, 177
China  18, 20, 29, 32, 58, 60, 81, 84, 86, 116, 117, 118, 119, 129, 144, 177, 194, 195
Christians  212, 226
Churchill, Winston - Prime Minister  83, 120, 130, 189, 221, 245
CIA  26, 39, 43, 52, 66, 69, 75, 79, 80, 81, 94, 131, 136, 155, 157, 164, 165, 168, 171, 177, 185, 186, 188, 201, 211, 213, 215, 253. *See also* Central Intelligence Agency
CIA's analysis  171
Citizen of the Year  199, 201
citizen soldiers  21, 35, 36, 37, 38, 61, 77, 88
Coalition  9, 24, 33, 55, 63, 71, 90, 208, 223, 224, 225
Coalition forces  24, 33, 63, 90
Coalition troops  71, 225
Coast Guard  113
Cold War  18, 22, 25, 26, 28, 30, 31, 54, 56, 58, 62, 69, 72, 79, 80, 82, 84, 87, 107, 112, 117, 118, 132, 140, 141, 142, 143, 144, 155, 161, 162,

165, 166, 179, 180, 183, 184, 192, 195, 204, 211, 212, 214, 215, 216, 228, 231, 240, 243, 245, 246, 249, 253
Coleman, Secretary  93
College Station  184, 206
Colorado Springs  45, 105
Commanders Emergency Response Program (CERP)  24, 33, 63, 90
Commission on the National Guard and Reserves  38
communications  20, 27, 47, 60, 86, 117, 132, 166, 194, 214, 215, 219, 220, 240
Communism  143
Communist  191, 192
Community College of the Air Force  204
Conference of European Armies  179
Congress  7, 22, 24, 25, 31, 32, 34, 38, 53, 54, 62, 64, 76, 79, 84, 88, 90, 95, 101, 102, 108, 109, 110, 126, 133, 151, 153, 160, 176, 209, 215, 217, 230, 231, 244, 247, 248, 250, 251
Constitution of the United States  101, 104, 109
constructive relationship  29
Container Security Initiative  113, 194
Continental Army  97
conventional forces  25, 131
Corder, Frank - Captain  121
CORDS  217, 218
Cornwallis - Lord  186
counter-narcotics  28, 115, 116, 231
credibility  8, 10, 57, 183
Crocker - Ambassador  132
Crystal Gateway Marriott Hotel  127
Cuban soldiers  82
Czechoslovakia  82, 211
Czech Republic  20, 29, 60, 86

D

Dallas  77, 78
Dallas Chamber of Commerce  77
Dallas-Fort Worth airport  78
Dawson, Joseph - Captain  122
Day - Colonel  146
D-Day  9, 120, 122, 129
Defense Budget  18, 58, 84
Defense Department. *See* Defense, Department of
Defense, Department of  8, 10, 11, 13, 16, 18, 31, 38, 42, 44, 58, 76, 77, 78, 84, 91, 101, 109, 137, 152, 158, 197, 211, 215, 217, 219, 225, 237, 241, 245, 247, 248, 250
democracies  29, 30, 114, 141, 143, 144, 175, 183, 193, 207
Denmark  28, 29
de-nuclearization process  118
Deny Flight  47
Department of Defense. *See* Defense, Department of
Department of State. *See* State, Department of
Desert Storm  47, 69, 124, 131, 155, 161
Devil Dogs  100
Dien Bien Phu  220, 245
Director General of the Foreign Service  241
Director of Central Intelligence  69, 79, 93, 113, 124, 139, 253
Distinguished Intelligence Medal  253
District of Columbia  210
DoD programs  88
Doolittle Raiders  105, 147
Doonesbury  185
Duke of Gloucester Street  94
Durbin - General  210

E

Eagleburger, Larry  25
Eastern Europe  26, 82, 184
East Timor  192
Eban, Abba - Israeli Foreign Minister  128, 169
Echo Company  133, 134
Edelman, Eric  53
education scholarships  238

Eeyore  139
Egypt  27, 69, 70, 71, 72, 73, 171, 172
Egyptian  69, 70
Ehlers, Walter  121
Eisenhower Administration  20
Eisenhower, Dwight - President/General  20, 60, 86, 157, 175, 242, 243
Ellis, Joe - Historian  140
English Channel  120
equality for women  27
Erhard, Ludwig - Chancellor of Germany  127
Esposito, Phil - Canadian hockey player  157
Estonia  28
Ethiopia  82
Europe  25, 26, 27, 28, 29, 60, 82, 86, 97, 114, 115, 117, 118, 122, 129, 140, 142, 143, 174, 179, 180, 181, 182, 183, 184, 212
European  20, 26, 27, 114, 141, 142, 157, 169, 179, 230
Evil Empire  143, 144, 211
extremism  8, 10, 27, 53, 54, 123, 125, 168, 173, 193, 246, 249
extremist networks  27, 81, 112, 175
extremist rhetoric  54

F

F-4  176
F-15  176
F-22  242
F-22A  20, 60, 86
Fallon - Admiral  49, 50, 123
Fallujah  96, 100, 133, 134
family housing  21, 61, 88
family units  21, 40, 61, 88
Fargo - Admiral  49
Federalists  141
Fifth Fleet  49, 222
First Cavarly  207
First Chief Directorate  164
First World War  129, 212. *See also* World War I

Fiscal Year 2008  18, 58, 84, 151, 152
Fitzwater, Marlin  185
flashpoints  18, 32, 58, 81, 84
Fleet Marines  129
Florida  123, 130
Flying Tigers  105, 147
Force Protection  23, 32, 63, 89
Force Space Command  176
Ford, Gerald - President  103, 143, 211
Foreign Service  132, 216, 219, 220, 241, 242, 244
Forrest Gump  102
Fort Belvoir  39
Fort Hood  206, 207, 208, 209
Fort Leonard Wood  202
Fort Myer  65, 154
Fort Riley  210
Forward Operating Base Tillman  40, 96
Founding Brothers  140
Founding Fathers  96, 102, 110
Fourth ID  207
Fouty, Byron - Private  150
France  120, 122, 141, 142
Franklin, Benjamin  186
Franks, Tommy - General  136
freedom of expression  27
free world  29, 76, 120, 122, 182, 191
French radicalism  141
From the Shadows: The Ultimate Insiders Story of Five Presidents and How They Won the Cold War  253
Fulda Gap  162, 179
Future Combat System  20, 60, 86

G

Gallipoli  129
Gates, Becky  7, 185, 187, 203, 254
Gates, Bill  210
Gates, Robert M. - Secretary of Defense  253
Gdansk  26, 143

GDP  18, 29, 31, 58, 79, 84, 182, 220.
    *See also* Gross Domestic
    Product
GE  95
General Staff College  160
Geneva  171
Georgetown University  220, 253
Georgia  93
Germany  25, 81, 82, 127, 145, 162, 179,
    184, 229, 242, 243
Germany, Imperial  229
Germany, Nazi  229
Getty, Paul J.  93
Giambastiani - Admiral  135, 136
Global Train and Equip  151, 250
Global War on Terror. *See* War on
    Terror
Global War on Terror Request  18, 22,
    58, 62, 84, 88, 151
Gorbachev, Mikhail - General Secre-
    tary (USSR)  102, 144, 165
Gordon - Sergeant  125
Grammy Award  237
Granada  245
Greater Killeen Chamber of Com-
    merce  206
Great Patriotic War  168
Greece  53, 104, 146
Grenada  124, 142, 215
Gromyko - Foreign Minister (USSR)
    164
Gross Domestic Product  18, 29, 31, 58.
    *See also* GDP
Ground Forces  19, 59, 85
Guadeloupe  179, 180
guerrillas  132, 161, 216
Gulf region  56, 69, 70, 71, 74, 116,
    224, 227, 228
Gulf Security Dialogue  227, 228
Gulf War  9, 47, 166, 184, 195
GWAT  251
GWOT  22, 23, 24, 32, 62, 63, 89, 90

# H

Habur gate  54

Haiti  142, 145, 161, 167, 243
Hamilton, Alexander  141, 187
Hamre, John  239
Hap's Place  105
Harker Heights High School  208
Harvard University  221, 245
Havel, Vaclav  97, 143
Hayes, Admiral  49
Heart of Texas Defense Alliance  206
hegemonic ambitions  28
Heidelberg  179
Heineman - Governor  176
Helsinki  26, 143, 211
Helsinki Accords  211
Helsinki conference  26, 143
Hepburn, Katharine  94
Herndon  99
Herzl, Theodor  175
Hezbollah  131, 226
Hezbollah-linked  131
Hickam, Homer  92
Higgins - Boatmaker  129
Higgins, William - Lieutenant Colonel
    131
Hilton, Paris  200
Hitchcock, Alfred  93
Holland - General  123
Holy Land  72, 228
Horst, Teltschik  25
hostages in Iran  69, 124
House Appropriations Committee  58
House Armed Services Committee  9,
    202, 230, 249
Hue  133, 154
humanitarian crises  113, 194
human terrain teams  244
Hurricane Isabel  99
Hurricane Katrina  38, 44, 46
Hussein, Saddam  113

# I

IISS  222
immigrant populations  27
Improved Explosive Devises (IEDs)
    23, 163

**272**

Inchon 133
Incirlik Air Base 54
India 27, 81, 112, 115, 192, 243
Indiana 202, 253
Indiana University 202, 253
Indonesia 27, 50, 112, 193
Information Agency, US (USIA) 80, 132, 215, 216, 219, 241, 244
insurgents 24, 33, 34, 55, 63, 64, 90, 132, 161, 162, 216, 223, 225, 231
International Institute for Strategic Studies 111
international legitimacy 28
International Security Assistance Force (ISAF) 54, 180, 183, 231
Internet 91, 108
Iran 10, 18, 28, 32, 55, 56, 58, 66, 69, 70, 71, 72, 81, 82, 84, 118, 124, 132, 139, 171, 172, 173, 175, 193, 215, 222, 225, 226, 227, 245
Iranian "moderates" 55
Iranian Revolutionary Government 55
Iraq 8, 9, 10, 11, 12, 13, 14, 15, 16, 17, 19, 22, 23, 24, 28, 30, 31, 32, 33, 35, 36, 37, 40, 42, 44, 47, 53, 54, 55, 56, 59, 62, 63, 67, 68, 70, 71, 72, 74, 78, 80, 81, 82, 83, 85, 87, 89, 90, 96, 97, 106, 111, 112, 113, 116, 124, 130, 131, 132, 133, 139, 140, 141, 142, 144, 145, 147, 151, 152, 153, 155, 159, 160, 161, 162, 166, 167, 171, 172, 173, 174, 175, 177, 186, 193, 195, 202, 204, 207, 213, 216, 217, 218, 219, 222, 223, 224, 225, 226, 227, 229, 230, 231, 235, 236, 237, 240, 241, 243, 244, 247, 250
Iraqi Security Forces 23, 33, 63, 71, 89, 152, 225
Iron Curtain 29, 82, 143, 170, 182
Iron Horse Division 207
ISAF 54, 180, 183, 231
Islamic 114, 141, 167, 174, 212, 225
Islamism 54

Israel 69, 72, 169, 170, 171, 172, 173, 175, 228
Israeli 70, 128, 169, 171, 175, 228
Italy, Fascist 229
Ivanov, Russian Minister of Defense 25

J

Jackson, Scoop 170, 175
Jamestown 95
Japan 29, 49, 50, 112, 115, 145, 191, 192, 193, 194, 195, 196, 229, 242
Japanese Self Defense Force 192
Japan, Imperial 229
Japan National Press Club 193
Jeffersonians 141
Jemaah Islamiyah 114
Jennifer Harris, Captain 42
Jerusalem 171, 175
Jewish Institute for National Security Affairs 169, 170
Jews 170, 175, 226
jihadist 10, 28, 48, 56, 112, 141, 172, 173, 185
Jimenez, Alex - Specialist 150
Johnson, Lyndon -President 127, 170
Joint Chiefs of Staff 38, 102, 118, 123, 129, 136, 150, 155, 215, 220, 242, 247
Joint staff 7
Joint Strike Fighter 20, 54, 60, 86
Jones, Jim - General, Supreme Allied Commander in Europe 182
Jordan 27, 172
JROC 136

K

Kabul 115, 218
Kagan, Donald - Historian 220, 245
Kandahar 28, 183
Kansas 7, 94, 200, 201, 202, 210, 211, 218, 233, 239, 245, 253
Kansas State 218, 239, 245
Kansas State University 202
KC-135 21, 61, 87

273

KC-X  21, 61, 87
Keating, Admiral  45, 46, 47, 49, 50, 51
Kelso, Dr.  93
Kennedy, John F. - President  92, 94, 188, 212, 243
KGB  164, 165
Khalilzad, Ambassador  14, 17
Khowst  232
Kipling, Rudyard  53
Kissinger, Henry - Secretary of State/National Security Advisor  128, 169, 181
Kitty Hawk  146
Kohl, Helmut - German Chancellor  25, 102
Kookaburra  147
Korea  18, 20, 29, 31, 32, 58, 60, 79, 81, 82, 84, 86, 111, 114, 115, 117, 118, 139, 145, 192, 193, 194, 215, 245
Korean Peninsula  191
Korean War  117, 246, 249
Kosovo  181, 192, 212
Kosygin, Aleksei - Premier (USSR)  170
Krulak, Victor - Marine 1st Lieutenant  129, 130, 132, 133
Kryuchkov, Vladimir - Head of KGB's First Chief Directorate (USSR)  164, 165
Kurdish  54
Kuwait  69, 78, 113, 142, 222

## L

Lackland  146
Laird, Melvin - Secretary of Defense  128
Land-Grant Colleges  253
Landon  210, 239
Landstuhl  158, 159
Lavrov - Foreign Minister (Russia)  166
LBJ ranch  127
Leavenworth  210
Lebanon  131

LeMay, Curtis  45, 105, 147
Leonardo da Vinci  146
Lincoln, Abraham - President  40, 94, 97, 99, 105, 170, 186, 188, 212
Lingle, General  49
Lippmann, Walter  188
Louis XVI  142

## M

MacDill  123
Macke, Admiral  49
Magnus, General  99, 127
major challenges  193
Maliki, Prime Minister  14, 17
Manama  222, 223
Manama Dialogue  222, 223
Marine Corps  10, 11, 12, 13, 15, 16, 19, 20, 21, 23, 32, 36, 37, 42, 59, 60, 61, 62, 75, 80, 81, 85, 86, 88, 89, 100, 101, 123, 127, 129, 133, 134, 209, 247, 250
Marine Corps Association  127
Mars  95
Marshall, George - General, Secretary of State  221
Marshall Plan, The  215
Martin Luther King  233, 234
Maryland  99
Maupin, "Matt" - Staff Sergeant  150
Mazar-e-Sharif  218
McCain, John - Senator  12, 15, 18
McGlothlin, Ryan  97
McHugh, Representative  230
McKiernan, David - General  97, 179
McKinney - Chancellor  206
Medal of Honor  121, 125
medical crises  113
Mediterranean  53
Meir, Golda - Prime Minister of Israel  128, 169
MIA  149
Middle East  8, 10, 14, 17, 26, 28, 46, 48, 53, 56, 70, 71, 72, 73, 81, 82, 83, 100, 114, 116, 144, 163, 170, 171, 172, 173, 175, 192, 193, 212,

222, 223, 224, 225, 226, 228, 229
Military Academy of the General Staff 164
military arena 243
military members 21, 61, 88
military modernization 18, 32, 58, 81, 84
Military Spouse Career Advancement Initiative 197
Military Technical Revolution 166
Minnesota National Guard 35
Missile Defense 20, 60, 86
Missouri 45, 202, 203, 204, 205, 254
Missouri National Guard Training Center 202
Mitchell, Billy 105, 147
Mitterand, MItterand 102
Modernization 20, 60, 86
Mongolia 112, 192
Morin, Minister 121
Morocco 27
Moro Islamic Liberation Front 114
Moscow 118, 157, 164, 165, 166, 195
Moseley, General 105, 146
Mossad 171
Moyers, Bill 127
MRAP 152, 235, 236
Mubarak, President 69, 70, 71
Mullen, Mike - Admiral - Chairman of the Joint Chief of Staff 99, 154, 158, 220, 230, 242
multi-ethnic Arab state 225
Multi National Forces 67
Munich 25, 26, 27, 55, 114, 118, 182
Munich security conference 114
Murtha, Chairman 58
Muskie, Ed 96
Muslim populations 114
Muslims 226
Muslim world 53

# N

NACCO Industries, Inc. 253
narcotics trafficking 165, 227
NASA 92
National Association of Rocketry 91
National Association of State Universities 253
National Defense Authorization Act 247, 250
National Eagle Scout Association 253
National Guard 10, 12, 15, 21, 35, 36, 37, 38, 44, 61, 81, 87, 88, 152, 202, 243
National Guard Bureau 35, 38
National Guard Empowerment Act 38
National Intelligence Distinguished Service Medal 253
National Intelligence Estimate 226
National Junior Leader Training program 199
National Military Establishment 215
National Security Act of 1947 211, 214, 245
National Security Adviser 55, 253
National Security Council 8, 26, 47, 52, 69, 179, 211, 225, 253
national-security issues 170
National Security Medal 253
National Strategy for Maritime Security 113, 194
National Training Center 161
NATO 8, 26, 27, 28, 29, 30, 31, 54, 129, 136, 179, 180, 181, 182, 183, 184, 192, 230, 231, 232
NATO Secretary General 29
Naval Academy 74, 99, 109, 135
Navy Flag Officers Conference 74
NCOs 162
Netherlands 28, 184
Neufeld - Speaker 210
New Mexico, USS 136
New Orleans 129
New York 71, 114, 193, 218
New York's World Trade Center 193
New York Times, The 218
New Zealand 115, 191
Nicaragua 136
Nichol, President 93

Nightstalkers  124
Nimitz, Chester  101
Nimitz, USS  100
Nitze, Paul - Ambassador  171
Nixon, President  52, 128, 169
Non-Military Assistance  24, 33, 63, 90
NORAD  45, 50
Normandy  121, 129, 168
North Africa  114, 242
North America  26, 27
North Carolina  233
NORTHCOM  45, 46, 47, 48
Northeast Asia  111, 118
Northern Command, U.S.  45, 46, 49
Northern Watch  47
North Korea  18, 20, 32, 58, 60, 81, 82, 84, 86, 114, 118, 139, 193, 194
Northwest  185, 199
nuclear ambitions  18, 32, 58, 81, 84, 194
nuclear program  56, 71, 172
Nunn, Sam  239

## O

O'Brian, Conan  210
O'Connor, Sandra Day - Justice  93, 139
Office of Strategic Services  215
Office of War Information  215
Offutt Air Force Base  176
Ogarkov - Marshal, Chief of the General Staff (USSR)  164, 166
Old Chicago  105
Olmert - Prime Minister (Israel)  171
Olson, Eric - Admiral  125
Olympic athletes  27
Omaha  120, 122, 176, 203
Operation Desert Storm  69, 124
Operation Eagle Claw  124
Operation Enduring Freedom  186
Operation Iraqi Freedom  47, 49, 131, 150, 186
Operation Urgent Fury  124
Origins of War  220, 245

Ottoman  54, 175
Ozal, Turgut  54

## P

Pace, Peter - General  7, 9, 41, 49, 100, 119, 123, 129, 133, 137, 150, 151, 154, 155
Pacific  46, 47, 49, 50, 51, 111, 129, 165, 191, 192, 194, 195, 199
Pacific community  195
Pacific Northwest  199
Pacific Rim  192
Pakistan  27, 40, 112
Palestinian  70, 175, 228
Panama  142
Parker Drilling Company, Inc.  253
Patriot missile programs  20, 60, 86
Patriquin, Travis - Captain  163
Pentagon Auditorium  7, 42
Pentagon Parade Field  149
Pentagon press corps  108
Pershing missiles  26, 180
Persian Gulf  175, 202, 212
Peru  157, 169
Petraeus, David - General  55, 132, 174, 247, 250
Phantom Canyon  105
Phi Beta Kappa  97
Philippines  112, 191, 193
Philmont  199
piracy  50, 194, 227
PKK  54
Point du Hoc  121
Pope, The  128
POW  149
Powell, Colin - General  234, 241
Powell, Don - Chairman of the A&M Board of Regents  185
Prague  183
Prague summit  183
Presidential Citizens Medal  253
Primakov, Evgeniy - Head of Russian Foreign Intelligence Service  165
Princeton  221

privatize 21, 40, 61, 88
Proliferation Security Initiative 112, 193
Protestant 212
Provincial Reconstruction Teams (PRT) 54, 115, 133, 167, 180, 218, 230, 232
Puller, Chesty 101
Purple Heart 208
Putin, President of Russia 25, 118, 164, 166, 168

## Q

Quadrennial Defense Review 75
Quality of Life 21, 61, 88

## R

Rabin, Yitzhak - Defense Minister/Prime Minister (Israel) 102, 171
Ralston, Joe - General 54
Ramadi 130, 162, 163
Ramadi, Voice of 130
RAND 244
Rangers 121
Ready First Brigade 162
Reagan Doctrine 40
Reagan, Ronald - President 40, 70, 143, 170, 187, 211, 222
Recapitalization 21, 61, 87
reconnaissance 20, 60, 86, 100, 166
recruiting 11, 13, 16, 21, 36, 61, 88, 135
religious toleration 26, 27
Rempt, Admiral 100
Renuart, General 45, 47
Republic of Korea 111, 192, 194
reserve component personnel 37
Reserves 10, 11, 12, 13, 15, 16, 21, 37, 38, 61, 88
reunification 82, 184
Revolutionary War 35, 36
Revolution in Military Affairs 166
Rice, Condoleezza - Secretary of State 9, 81, 195, 241
Rickenbacker, Eddie 105, 147

Riga Summit 29
Ring Road 115
Ritchie, Steve 105
Riyadh 132
Rochelle, Michael - Lieutenant General 234
Rogers, Will 53, 96
Romania 28
Roosevelt, Theodore - President 109, 188, 214, 221
ROTC 96, 210, 217
Rudder, James Earl - Lieutenant Colonel 121
Rumsfeld, Donald 7
Russell, Richard - Senator 93
Russia 18, 25, 27, 29, 32, 58, 60, 81, 84, 86, 113, 116, 118, 144, 157, 164, 166, 168, 194, 195, 212
Russian Foreign Intelligence Service 165

## S

SAC Headquarters 203
Sadat, Anwar - President 69, 102
Saddam-era 223
Sadr City 207
Saigon 117
SALT II 164
San Francisco 191
San Francisco conference 191
SARS 113, 194
Scharf, Patricia 150
Scholtes - Major General 126
Schoomaker, Pete - General 65, 67, 80, 123, 124
Schultz - Secretary 240
Schwartz - General 123
Scout Councils 199
Scout Oath 201
Scowcroft Award 184
Scowcroft, Brent 52, 184, 185
seaborne threats 227
SEAL 126
Second World War 193, 243. *See also* World War II

Secretary of State  9, 25, 81, 96, 128, 166, 169, 181, 195, 219, 221, 228
Secretary Rice. *See* Rice, Condoleezza - Secretary of State
sectarian conflict  28, 71, 212, 240, 246, 249
sectarian divides  225
Section 1206  22, 62, 87, 151
Security Consultative Meeting  192
Security Council. *See* U.N. Security Council
Selfridge, Thomas - Lieutenant  106
Senate  7, 8, 9, 12, 15, 18, 31, 42, 84, 127, 139, 151, 202, 231, 246
Senate Appropriations Committee  84, 151
Senate Armed Services Committee  9, 12, 15, 246
Senate Youth Program  42
Serdyukov - Defense Minister (Russia)  166
Seville  26, 28, 31
Shamir - Prime Minister (Israel)  171
Shanghai  129
Shangri-La  111, 113, 119
Shangri La Dialogue  113, 222
Sharansky, Natan  170
Sharm al-Sheikh  71
Shaw, George Barnard  52, 94, 105, 210
Shia  9, 212, 225
Shin Bet  171
Shipbuilding  20, 60, 86
Shoemaker. Bob - General  206
Shughart- Sergeant  125
Sills, Beverly  93
Silver Buffalo Award  201
Simpson, Alan - Senator  127, 206, 239
Singapore  111, 112, 193, 222
Singh, Prime Minister  112, 192
Six-Party process  118
Six-Party-Talks  194
Skelton, Ike - Congressman  202, 205
smuggling  227
SOCOM  123

soft power  216, 232, 242
Somalia  27, 125, 161, 167, 216, 243
Sophia University  191
South Africa  140
Southeast Asia  82, 114, 124
Southern California  161
Southern Watch  47
South Korea  29, 115, 117, 145, 192
South Vietnam  218
Southwest Asia  116, 118, 212
Soviet Navy  100
Soviet Union  26, 56, 72, 82, 92, 113, 115, 140, 141, 142, 143, 147, 164, 167, 168, 170, 172, 174, 179, 184, 195, 203, 205, 211, 214, 215, 228, 229
Space and Naval Warfare Center (SPAWAR)  235
Space Capabilities  20, 60, 86
Space programs  20, 60, 86
Spain  27
Spaso House  165
Spears, Britney  200
Special Envoy  54
Special Operations Command  66, 123, 124
Special Operations Forces  19, 59, 85
Spratt - Congressman  146
Spring Offensive  180
Sputnik  92, 211
SS-20  179
Stalin, Joseph  142
Stanford  97
Stanton, Edwin - Secretary of War  186
State Department. *See* State, Department of
State, Department of  31, 43, 47, 80, 81, 132, 151, 165, 216, 217, 219, 220, 232, 241, 242, 243, 244
State of the Union  46, 85, 243
Stearns - Congressman  146
Stennis, USS  100
Stephenson, Sir William - Author  213
STRATCOM  176, 177

Strategic Air Command  176
strategic crossroads  29, 194
strategic environment  18, 27, 28, 29, 32, 46, 57, 58, 84, 246, 249
Strategic Investments  20, 60, 86
Strauss, Bob  169
submarine  100, 136, 158, 165
Sudetenland  220, 245
Sunni  9, 212, 225
Sun Tzu  59, 85
Sununu, John - Chief of Staff  185
Supplemental  18, 22, 23, 24, 31, 32, 33, 62, 88
Supplemental Appropriation Request  18, 31
Supreme Court  52
surge  9, 75, 162, 223, 240, 241
surveillance  20, 60, 86
Switzerland  171
Syria  55, 71

## T

Taliban  23, 28, 33, 180, 218, 230, 231, 244
Tallil  96
Tampa  123
Tampa Convention Center  123
Tarawa  133
Task Force 160  124
Team America Rocketry Challenge  91
Tehran  56, 71, 72, 172, 220, 227, 228, 245
Terrorist attacks  112
terrorists  24, 34, 55, 64, 87, 88, 90, 100, 112, 114, 115, 131, 132, 161, 167, 180, 182, 185, 193, 201, 207, 212, 216, 223, 225
Texas  7, 42, 43, 49, 74, 77, 78, 91, 93, 94, 95, 99, 105, 121, 176, 185, 186, 188, 203, 206, 208, 218, 239, 244, 253
Texas A&M University  7, 42, 43, 44, 49, 74, 77, 78, 91, 93, 94, 95, 99, 105, 121, 176, 185, 186, 188, 206, 218, 239, 244, 253

Thatcher, Margaret  102
The Fidelity Funds  253
The First Team  207
The National Security Act  243
The Way Forward in Iraq  202
Third Reich  142
Third World War  167
Tidal Basin  196
Tikrit  207
Titanic  52
Tokyo  191, 193
Train and Equip Authorities  22, 62, 87
trans-Atlantic partnership  183
Transforming Land Forces in the 21st Century.  179
Tripler  158
Trudeau, Gary - Cartoonist  185
Truman Doctrine  53
Truman, Harry - President  53, 57, 77, 170, 191, 195, 204, 211, 214, 237, 239
Tunisia  27
Turkey  53, 54, 55, 57
Turkish- American relationship  57
Tuskegee Airmen  105
TV  108
twin towers  114

## U

under-funding  80, 160
United Kingdom  27, 28, 29, 60, 86, 131, 239
United Nations  112, 117, 195
United Nations Security Council  195
United States  7, 8, 10, 14, 17, 18, 19, 21, 25, 27, 29, 31, 40, 42, 43, 48, 50, 54, 55, 56, 57, 58, 59, 61, 65, 67, 69, 70, 72, 73, 79, 80, 81, 83, 84, 85, 88, 92, 94, 99, 100, 101, 104, 106, 109, 110, 111, 112, 113, 116, 117, 118, 120, 130, 131, 132, 137, 139, 140, 142, 143, 145, 146, 157, 161, 165, 169, 173, 175, 192, 193, 196, 206, 207, 208, 211, 212, 213, 215, 216, 217, 219, 221,

222, 223, 224, 225, 226, 227, 228, 229, 230, 232, 233, 234, 235, 243, 244, 245
United States Agency for International Development 215, 216
United States Congress 7, 25, 54
United States intelligence community 225
United States Naval Academy 99
UN peace operations 192
U.N. Security Council 118
USAID 217, 220, 232
Usama bin Laden 80
U.S.-China relationship 119
U.S.-China Strategic Economic Dialogue 119
U.S.-Egypt relationship 69
U.S. forces 9, 14, 17, 23, 33, 62, 63, 89, 160, 192, 223
U.S. Government 20, 60, 86
USO 237
U.S. Pacific Command 49
U.S.S.R 144
U.S. supremacy 20, 61, 87
Ustinov - Defense Minister (USSR) 164
U.S. Transportation Command 21, 61, 87
U.S. troops 9, 14, 17, 183, 217, 231, 235

V

Vandenberg Air Force Base 204
Vienna 164
Vietnam 18, 31, 58, 79, 80, 81, 82, 84, 95, 117, 129, 135, 150, 155, 160, 166, 189, 205, 211, 215, 216, 217, 218, 241, 243, 245, 246, 249
Vietnam War 129, 155, 160, 166, 241
VIP 146
Virgil 92
Virginia 65, 91, 92, 93, 95, 140, 145
Virginia Tech 95

W

Walesa, Lech 102, 143
Wal-Mart 95, 119
Walter Reed Hospital 40, 76, 78, 96, 102, 110, 153, 158, 202
War Funding Request 250
Warner - Senator 127
War on Terror 18, 19, 22, 31, 32, 36, 58, 59, 62, 74, 84, 85, 87, 88, 150, 151, 152, 162, 207, 216, 250
Warsaw Pact 27, 183
Washington 7, 9, 12, 14, 15, 17, 18, 30, 31, 35, 36, 39, 42, 43, 49, 52, 53, 58, 65, 71, 77, 79, 81, 84, 85, 97, 102, 110, 119, 123, 127, 129, 141, 142, 145, 151, 157, 161, 164, 165, 176, 185, 186, 187, 188, 196, 199, 200, 206, 210, 211, 230, 233, 237, 239, 241, 246, 249
Washington D.C 39, 233
Washington, George 30, 35, 85, 97, 102, 110, 141, 145, 187
Washington Monthly 188
Washington Post 161
Watergate 52
Welch, Larry - General 199
Wenceslas Square 26
Western democracies 143
Western European 26
Western Iraq 116
West Virginia 92
White House 9, 52, 66, 102, 124, 127, 132, 215, 239, 253
Whiteman Air Force Base 39, 45, 106, 146, 155, 202, 254
Wichita 211, 233
Wichita High School East 211
William and Mary, College of 42, 93, 139, 140, 253
Williamsburg 93, 94, 139
Wilson- Congresswoman 146
Wilson, Woodrow - President 140
Winter, Secretary 99
Wisner, Frank - Deputy Executive Secretary 171
Woolworth 233
Worden Field 99

World Forum on the Future of Democracy  139
World Trade Center  114, 193
World War I  113, 212. *See also* First World War
World War II  66, 79, 119, 157, 168, 175, 212, 213, 214, 215, 235, 242, 244. *See also* Second World War
Wren Building  94
Wright, Orville  106, 146, 147
Wynne - Secretary  105, 146

# Y

Yeltsin, Boris - President (Russia)  102, 165
Young, Bill - Congressman  123
Young, Congressman  58
YouTube  200
Yugoslavia  113, 212

# Z

Zarqawi  207
Zembiec, Douglas - Captain  133, 134
Zionism  175

www.ingramcontent.com/pod-product-compliance
Lightning Source LLC
LaVergne TN
LVHW041611070426
835507LV00008B/187